INTERNATIONAL DEVELOPMENT IN FOCUS

Unlocking the Lower Skies

The Costs and Benefits of Deploying Drones across Use Cases in East Africa

AIGA STOKENBERGA AND MARIA CATALINA OCHOA

Contents

Maps

Photographs

Tables

Foreword

The transport sector is a critical component of economic recovery from the COVID-19 (coronavirus) pandemic, yet transport and logistics costs in Africa remain high, especially for reaching remote communities. Only 34 percent of Africans live within 2 kilometers of an all-weather road in rural areas compared with more than 90 percent in East Asia. The consequence is that hard-to-reach communities face higher costs for goods and long wait times for deliveries; rural enterprises suffer from lower productivity, and rural citizens have fewer opportunities to break the cycle of poverty. Africa needs to spend billions on new transport infrastructure as well as operations and maintenance just to sustain current levels of development. In addition, roads are simply not safe. Africa has 2 percent of the world's road vehicles and 16 percent of road fatalities. Therefore, we need to rethink how to better deliver mobility, such that the return on infrastructure is both greater, safer, and itself more resilient.

New technologies are transforming transportation, and electric autonomous drones epitomize the immediate leapfrogging opportunity for emerging markets. Improvements in electrification, autonomous operations, and new business models for ride sharing are disrupting transport markets and opening up new opportunities. Electric motors and propulsion technologies are, today, cheaper to run and maintain than gasoline engines. The cost of batteries for electricity storage has been dropping for the past decade, and this trend is accelerating while energy density has been increasing. Progress in autonomous technologies also means vehicles can become cleaner, cheaper, and safer to operate.

The use of clean and safe autonomous electric technologies is growing rapidly in the humanitarian sector, and Africa is leading the way. Lower-cost aircraft that are much cheaper to operate and simpler to maintain create the possibility of establishing delivery routes to smaller airfields and hard-to-reach communities or of operating in more hazardous conditions than manned systems. These systems can also benefit from lower capital requirements given that drone ports are expected to cost about as much as a petrol station compared with much more expensive airports. The current opportunity is to rethink the infrastructure needs for a new kind of supply chain, one that has greater reach and resilience. Malawi and Sierra Leone have set up humanitarian corridors where they are learning about and testing new technologies and services. Rwanda has pioneered countrywide lifesaving blood deliveries, putting a majority of the country's citizens within 20 minutes' reach of critical medicine. Experience in the Democratic Republic of Congo has demonstrated that drone technology can cut vaccine delivery time from roughly 3 hours to 20 minutes one way.

This report shows that drones can already bridge some of these gaps in a cost-effective way, delivering significant human and societal benefits. In many contexts, drones are close to or at cost parity with traditional transport modes, such as trucks and motorcycles, in medical goods delivery and are an economically attractive alternative to manned aircraft in humanitarian food aid delivery in conflict situations. Although they are not yet cost competitive with land transport for delivery of large volumes of goods, such as food staples, electric cargo drones offer cost savings in indirect ways, such as by reducing inventory stock needs and uncertainty in delivery times when roads are impassable for significant parts of the year. Moreover, drones offer cost and flexibility advantages compared with manned aircraft in mapping, disaster risk assessment, and agricultural applications. This book sheds light on the potential of drones to save lives by getting essential medical goods and blood for transfusion to patients who would otherwise die, and allowing people to be diagnosed and start treatment sooner. Although more research is needed, particularly to understand the potential of drones as a service, there is a clear opportunity for drones to contribute to the development of advanced digital skills in Africa.

The potential cost and flexibility advantages of drones are magnified in the context of COVID-19. The global health and economic crises show the vulnerability of essential goods supply chains across not only Sub-Saharan Africa but also many developed countries, whether from climate events or pandemics. Disruption from COVID-19 resulted in scarcity of medical products, food, hospital beds, and personnel. In the current environment of severe in-person travel restrictions, drones can step in for critical activities such as risk mapping, crop inspections, infrastructure inspections, and monitoring. Countries that have proactively built drone delivery networks are better prepared to reach the last mile in remote areas with critical personal protective equipment delivery, lab sample collection, and vaccine distribution.

Drones can accelerate Africa's digital economy. Cargo and supply chain drones can serve businesses and communities that are digitally online, but physically isolated, with lower-cost or faster and safer on-demand deliveries. From vaccines and medicines to spare parts, from seeds to insecticides, and from documents to cash on demand, drones can boost digital economic opportunities. The World Bank is committed to supporting disruptive technologies with high potential for significant positive social impacts. Drones are one of the most promising and important of these technologies. I am sure that this report will greatly contribute to realizing this potential and will inspire joint work by development partners, regulators, and industry to unleash the unlimited potential of the lower skies as a green, safe, and inclusive resource for growth.

Hafez M. H. Ghanem
Vice President, Eastern and Southern Africa Region
The World Bank

Acknowledgments

This report was prepared by a team led by Aiga Stokenberga (transport economist, Transport—West Africa and Nigeria) and Maria Catalina Ochoa (senior urban transport specialist, Transport—East Africa). Overall managerial support was provided by Franz Drees-Gross (regional director, Infrastructure—Latin America and the Caribbean), Ben Eijbergen (practice manager, Transport—East Africa), and Riccardo Puliti (regional director, Infrastructure—West Africa).

Valuable data inputs and support for the report were provided by the following World Bank staff and consultants: Edward Anderson, Sarah Antos, Rogers Ayiko, Tatiana Daza, Vivien Deparday, Yasser El-Gammal, Cristiano Giovando, Jessica Gosling-Goldsmith, Richard Martin Humphreys, Yerassyl Kalikhan, Toni Lee Kuguru, Julia Mensah, George Mulamula, Patrice Mwitende, Li Qu, Tojoarofenitra Ramanankirahina, Robert Reid, Keiko Saito, and Hadia Samaha.

Peer review of the study was provided by Niels B. Holm-Nielsen (practice manager, Global Facility for Disaster Reduction and Recovery), Son Nam Nguyen (lead health specialist, Health, Nutrition, and Population—Middle East and North Africa), and Charles E. Schlumberger (lead air transport specialist, Transport—Global Knowledge).

The team would like to express gratitude for the external inputs and comments provided by Juma Ikombola, Matiko Machagge, Catherine Nyinondi, Naomi Printz, Cary Spisak, and Amani Thomas at John Snow, Inc. (JSI); Gabriella Ailstock, Olivier Defawe, and Luciana Maxim at VillageReach; Tedla Damte and Tautvydas Juskauskas at UNICEF; Kate Langwe, Collings W. Mfungwe, and Charles Mwansambo at the Ministry of Health and Population of Malawi; Simon Grandjean-Lapierre at the Global Health Institute, Stony Brook University; Astrid Knoblauch at the Drone Observed Therapy System in Remote Madagascar (DrOTS) Project; Oleg Aleksandrov, Takako Kaneda, and Sumalee Steruphansen at the World Food Programme; Frederick Mbuya at Uhurulabs; Leka Tingitana at Flying Labs Tanzania; Kush Gadhia at Astral Aerial Solutions Ltd.; Eric Rutayisire at Charis; Bradley Friedman, Jordan Litter, Zahra Nensi, and Manraj Singh at Deloitte; and Scott Dubin, Ashley Greve, and Ryan Triche at Chemonics.

This work was made possible by financial support from the World Bank Transport Practice; the United Kingdom Foreign, Commonwealth and Development Office through the Tanzania Corridors for Growth Trust Fund; and the Ministry of Land, Infrastructure and Transport (MOLIT) of the Republic of Korea through the Korean Institute of Aviation Safety Technology (KIAST); as well as technical support from the African Drone Forum and its network of partners.

About the Authors

Maria Catalina Ochoa is a senior transport specialist at the World Bank working at the intersection of transport, data, and technology. She manages several urban transport initiatives in the East Africa region, while also serving as global lead for innovations platforms and disruptive technology in the sector. Maria Catalina leads the African Drone Forum and several other drone initiatives at the World Bank. Before moving to East Africa, she led the World Bank urban transport portfolio in Argentina and Mexico and worked in operations in Latin America and Asia. Maria Catalina has substantial experience in the tech and start-up space. With Amazon she worked on their supply chain and logistics strategy, and at Uber she was strategy manager and general manager in Latin America and Africa and the Middle East. Maria Catalina is a recognized voice in the transport community and has published on transport and open data, disruptive technology, drones, jobs accessibility, clean technology, logistics, climate change, gender, and other topics. She is an engineer with an MSc in transport engineering and an MCP in city and regional planning from the University of California at Berkeley, and an MBA from INSEAD.

Aiga Stokenberga is an economist in the Transport Global Practice of the World Bank, where she leads economic and spatial analyses that inform urban transport and regional transport corridor planning strategies in Latin America, South Asia, and Sub-Saharan Africa. She previously worked in the fields of sustainable energy, logistics, and trade integration at the World Bank, the World Resources Institute, and the Ross Center for Sustainable Cities. Her published research spans the fields of urban and infrastructure economics and transport planning. Aiga holds an MA in international energy policy from Johns Hopkins University School of Advanced International Studies and a PhD in environment and resources, with a focus on urban land use, from Stanford University.

Executive Summary

INTRODUCTION

The current gaps in road transport infrastructure in East Africa and Sub-Saharan Africa more broadly are vast, amounting to billions of dollars annually. These gaps present enormous challenges to meeting the Sustainable Development Goals, from health to agricultural productivity to food security, and can only be expected to be magnified by the consequences of climate change.

As demonstrated by the global economic and health crisis unfolding in 2020 and 2021, essential goods supply chains across Sub-Saharan Africa as well as in many developed countries are highly vulnerable to both climatic events and health pandemics. The COVID-19 (coronavirus) experience shows that traditional supply chains can be substantially disrupted by a pandemic, creating scarcity of medical products, food, hospital beds, and personnel. Although the key disruptions in health supply chains caused by COVID-19 have mostly resulted from factors such as increased global demand, restrictions on manufacturing and support industries, and bans on exporting selected essential health commodities by some source countries—rather than from constraints on the physical delivery of health commodities from warehouses and depots to facilities—unmanned aerial vehicle (UAV) technology has a growing role to play in more effectively managing inventories, protecting people from contamination, and delivering critical goods such as medical products, lab samples, vaccines, and food in a way that minimizes person-to-person contact. This realization provides additional strong impetus for seeking complementary transport solutions that can fill in the gaps and build more resilient supply chains.

This report explores the economic and broader societal rationale for introducing UAV, or "drone," technologies to complement current transport and logistics systems in several use cases in the context of East Africa. The specific use cases examined include medical goods deliveries, food aid delivery, land mapping and risk assessment, agriculture, and transport and energy infrastructure inspection. The first two of these correspond to cases in which physical deliveries of goods are performed using UAVs; the latter three focus on the role of UAVs mostly as a technology for image and information gathering and as an assessment tool. Across these applications, the case for using UAVs is examined within the context of the so-called logistics objectives—total operating costs, speed, availability, and flexibility—as well as the human, or societal, objectives, which may include increased access to health care, lives saved, and increased environmental sustainability, among others.

MEDICAL USE CASE

As more low- and middle-income countries explore opportunities to improve efficiency and performance in their public health supply chains and diagnostics networks, they face myriad choices about how best to use UAVs to improve product availability, public health outcomes, and reaching the last mile. UAVs have been implemented in several different contexts, proving that they can certainly reach the last mile with urgent health commodities. As public health supply chain managers move beyond pilots and toward implementation, incorporating UAVs as one type of vehicle in their overall fleet, it is important to identify which product groups or flows may be the best candidates for UAVs, and to compare the cost of UAVs with the cost of current transport modes.

The high-level findings from this analysis are that, if examining commodity categories individually, and looking exclusively at costs, delivery with UAVs in general is still more expensive. However, the value of drones tends to increase with higher density of health facilities within the range of unmanned aircraft systems (UASs[1]); greater difficulty of accessing the facilities by road; higher financial value, scarcity, or health value (lifesaving nature) of the medical goods; unpredictability of demand at the level of individual facilities; and shorter shelf life and greater difficulty of storing the medical goods at health facilities. As part of this study, in-depth data collection and interviews with health sector workers were conducted in Ukerewe District, including the district's islands in Lake Victoria, which forms part of the Mwanza region of Tanzania, to identify optimal use cases in the medical goods delivery space to demonstrate the specific logistical challenges and problems in the current system to which a UAS could effectively respond. Categories with very high quantities of products result in relatively low costs per flight, but significantly more flights and UAVs are required to deliver the same quantities as the current delivery system; the net result is higher costs. Small volumes (such as in the case of lifesaving items) and infrequent demand result in few flights with high costs per flight. Public transport is widely used for medical goods transport and is relatively inexpensive, keeping current delivery costs low.

Although still higher cost than current delivery systems, the most cost-effective use case examples include the transport of laboratory samples to selected destinations (associated with the lowest percentage increase in transport costs compared with the present) and delivery of lifesaving items and blood to selected destinations (associated with the lowest dollar amount increase in transport costs). UAVs are most cost-effective on routes to distant facilities on the small islands and between the District Hospital in Nansio (on the main island) and Mwanza.

However, layering use cases can provide efficiencies and cost savings by allocating fixed costs across a greater number of flights and maximizing drone capacity and time utilization. The base case absorbs fixed costs and start-up capital costs, and additional layered use cases incur incremental operations costs. Increasing the number of flights per UAV reduces overall per flight costs, and combining products per flight maximizes capacity usage (in liters or kilograms) of the vehicle. The cost-effectiveness of a UAV can be increased by maximizing the cargo capacity and time utilization in all directions, which case layering contributes to. For routine deliveries, cost-effectiveness can also be increased by using "milk runs" (visiting multiple facilities in one outing) instead of hub-and-spoke routes (going out and back between Nansio District Hospital and each facility) because excess capacity is used by including cargo for more than one facility, and total mileage traveled is reduced. In the Ukerewe context, opportunities for UAV transport cost reduction lie in layering lifesaving item and rabies vaccine deliveries on top of laboratory sample and blood deliveries that make up the base cases. As a result of such layering, the time utilization of a given UAV could be increased by 60–80 percent, leading to average transport cost savings per flight of about 30–40 percent. It should be noted, however, that, given the estimated time required to complete one flight, including preparation, packing, unpacking, and flying, a single UAV can be expected to perform no more than 1,200–1,600 flights per year; thus, the layering of use cases only provides cost efficiencies up to this ceiling; additional flights above the ceiling would require capital investment in additional UAVs.

Cost sensitivity analysis for a hypothetical "East Africa base case" demonstrates that any savings in transport costs for medical goods using UAVs as compared with traditional transport modes are mostly driven by drone vendor pricing structures and drone costs, ground transportation costs, and the level of demand:

- *Drone vendors vary greatly in their pricing and drone capabilities.* The capital cost of one drone may be nearly 10 times the price of another, so to be cost competitive, the cheapest vendor that meets the required specifications should be chosen.
- *Leasing (that is, the drones-as-a-service model[2]) provides a much more cost-competitive option compared with buying.* Leasing not only often results in lower overall costs, but large capital investment requirements are avoided and cost stability is increased compared with owning drones. In the hypothetical East Africa case analysis, leasing is the only cost-competitive option across all scenarios.
- *Because drones cannot compete on price with cheap or free transportation (including public transit), site selection is critical to a cost-effective drone program*; drones must be used where ground transportation is currently expensive.
- *Drones scale well because savings add up faster than costs.* When the population of the area to be served increases by a factor of 10, traditional ground transport costs of medical goods (by truck or motorcycle) increase by roughly six times, whereas drone costs only double. This relationship enables large savings to be realized from not using ground transportation, thereby offsetting the costs of the drones. However, having a prohibitively expensive drone or cheap or free ground transportation will prevent drones from ever reaching a cost-competitive scale, no matter how many trips the drones travel.
- *Costs for the infrastructure necessary to support drones, such as the costs of a drone port, can easily prevent drone programs from becoming cost competitive, regardless of demand levels or drone vendor specifications.* Offsetting high infrastructure costs by either leasing the drone or selecting the drone with the cheapest capital investment is critical to being cost competitive.

The Ukerewe case study suggests that the capital cost of a UAV that would bring drone usage to cost parity with current road- and boat-based transport is approximately $19,000. This estimate is significantly less than the $75,000 capital cost per UAV assumed throughout the baseline scenario cost analysis, but it is not impossible to reach given the speed with which the technology is developing, including in research and development facilities in emerging markets such as China. However, it should be noted that the estimated capital cost parity assumes several specific UAV technical capabilities, such as vertical takeoff, cargo capacity of at least 12 liters (4 kilograms), and a range of 100 kilometers (km), which the cheaper UAV would also have to meet. The estimated operating cost parity is extremely low for vaccine deliveries, at less than $0.01 per liter per km, regardless of the destination type. The estimated parity is by far the highest for lab sample pickup from the small island facilities and the distant big island facilities, at about $0.68 per liter per km. The scenarios modeled in this analysis did not include the up-front costs of procuring, importing, and setting up the UAS. These fixed costs should be taken into account if budgets are developed for UAS deployment.

The drone-as-a-service model, albeit still relatively new, provides complementary insights into cost parity results. The Malawi-focused cost model examining scenarios assuming the drone-as-a-service model finds that cost parity—or maximum cost-competitive monthly all-in drone-as-a-service price—is approximately $2,000, equivalent to the savings obtained from not using ground transportation. The case study also highlights the importance of the level of demand in determining the maximum price at which UAV services would be cost competitive with ground transportation. When using the Malawi data to calibrate the hypothetical East Africa base case model, it can be observed that if the number of monthly cases of an illness that requires lifesaving medicines, blood transfusion, or a similarly urgent intervention were to quadruple, the parity monthly drone-as-a-service cost would more than double, to $5,000.

Across most of the existing analyses, the main drivers influencing the relative cost advantages of UAV-based versus road transport–based medical deliveries are found to be vehicle and fuel costs. Other studies of the medical use case that have been implemented in the region report a range of conclusions, with some finding the transportation costs of a UAS higher than those associated with land-based transport and others suggesting that under specific assumptions, UASs are cost competitive.

From the perspective of public decision-makers or international donor organizations, the cost-effectiveness of UAVs cannot be analyzed without looking at the public health benefits, which may be substantial. In Rwanda, where the drone program has been operational for several years, stock-out rates have decreased to zero, and availability of rare blood products is estimated to have increased by 175 percent. In Madagascar, although the tuberculosis-focused drone pilot program was assessed as unaffordable from the government's perspective, as measured by the incremental transport costs per disability-adjusted life year, the health impacts of the trials were impressive, with tuberculosis diagnosis rates, and, subsequently, enrollment rates in treatment programs, doubling. In Ukerewe, the analysis conducted for this study using a decision-tree model suggests that the deployment of drones to deliver emergency blood for transfusion would reduce the number of deaths by almost 80 percent compared with traditional transport. Drones would achieve this level of success by reducing the travel time either from nearby health care facilities or from the National Blood Transfusion Services Laboratory Center located in Mwanza as well as by maintaining product quality. Evidence from the study shows that more than 10 percent of blood that is collected by health facilities is currently wasted because of hemolysis of samples that use traditional transport to Mwanza for screening. If drones were used to transport blood samples, wastage would be minimized and the need to borrow from nearby facilities or make requests from the blood center in Mwanza would be reduced.

The use of UAVs can also help reduce the longer-term negative health impacts stemming from the current diversion of health care workers' time to medical goods transport, although the extent to which health workers are directly involved in the transport of health commodities varies by country. *Public transport is an inexpensive mode of transport across Tanzania and is also used widely for transport of public health commodities in other countries in the region, such as Malawi.* Thus, from a cost perspective alone, it is difficult for UAVs to compete with public transport. However, public transport schedules can be limited and generally inflexible in meeting urgent needs, limiting the health center's ability to respond, which may result in significant impacts on patients. Although the Ukerewe analysis includes the cost of salaries and per diem in transport costs where applicable, it does not explicitly quantify the opportunity cost of health care workers' traveling by public transport to pick up the needed medicines to serve patients. The field research in Ukerewe confirmed that health care workers are tasked with the delivery and collection of goods. For facilities near the collection points the impact is likely negligible, but for distant facilities or for those with limited personnel, travel time requirements can be significant, effectively replacing time serving patients and leading to tangible health impacts over the long term that may or may not be possible to quantify.

Existing UAV impact assessments suggest the need to mix and match modes to serve needs; the trade-offs are going to be positive only in particular niches. In other words, the transport technology needs to be matched with system characteristics: accessibility and distance, product weight and volume, product value, and urgency of need. Despite some encouraging results from the existing efforts to quantify the advantages of UAVs through pilot operations in the medical goods delivery use case, there is still insufficient long-term, ex post data with which to accurately assess the comparative advantage of UAVs, either from the cost or the health impacts perspective. More in-depth research is needed on how transportation systems could include *both* land transport and UAVs to best take advantage of the benefits of each transport mode. Additional scenarios should focus on overall system optimization to define concrete implementation plans. Other specific areas for further research, in the Ukerewe context and elsewhere in the East Africa region, include a cost comparison between current transport modes and the use of UAV transport as a service (rather than purchasing UAVs); an assessment of how the current modes would compare with heavy-lift UAVs; and the potential advantages of a truck-UAV hybrid distribution model for routine products. Finally, future research could expand the health-impacts assessment of UAVs by studying health-seeking behavior resulting from increased availability of health products.

OTHER USE CASES

In the food aid delivery use case, large cargo drones may already be able to offer cost advantages on a per ton per km basis compared with regular manned aircraft, as illustrated by the analysis of data pertaining to the World Food Programme's (WFP's) operations in South Sudan. The most feasible scenarios in which drones may provide overall cost advantages compared with manned airplanes, however, could be in those cases, such as during true emergency conditions, when relatively small quantities of food would need to be delivered to individual destinations and therefore a small cargo plane would be the real counterfactual to drones. Moreover, drones may provide a practical advantage—or be the only practically feasible solution—in scenarios in which, because of climate or security conditions, a manned cargo plane would not be able to physically reach its airdrop destination and safely return. In the case study of food deliveries to hard-to-reach districts in Rwanda as part of the WFP–managed Home-Grown School Feeding program, where trucking is the counterfactual, drones do not appear to provide advantages in *direct up-front* transport costs, even for last-mile deliveries. However, drones can be a solution for ensuring timely delivery of the much-needed food items to the target schools and potentially reduce the delivery time uncertainty–related food inventory costs at the school level. Overall, in East Africa and elsewhere, there are still few cases of actual, even pilot-based, delivery trials in the cargo sector, and it should be expected that the costs of the vehicles and equipment will decrease significantly in the future.

Drone use in use cases such as mapping, risk assessment, and agriculture in East Africa is relatively more common than cargo drone operations, and existing pilot initiatives have delivered impressive results for speed and quality (precision). Drones have also been shown to be cost competitive with traditional aircraft, even if not with very low-cost, crowdsourcing-based data-collection methods, and offer practical advantages in conditions after natural disasters such as cyclones when traditional manned aircraft are not able to capture high-quality imagery because of the cloud cover. Compared with satellite imagery–based data-collection methods, drones appear cost competitive in contexts in which high-precision maps need to be produced on short notice. However, the use of drones in the mapping and risk and postdisaster needs assessment use cases remains only partially, or not at all, integrated into government systems, which to some extent is a function of the funding of the programs by external organizations. At least partly because of the significant know-how and skills requirements, which in some cases may be an even greater barrier to drone adoption in the disaster risk management space than their cost, across East Africa, the mapping and risk assessment initiatives that have used drone technology have tended to be contracted to private firms. An exception is the Zanzibar Mapping Initiative in which significant investment was made to build local technical capacity to ensure continuity. Most companies operating in the region that integrate drone imagery with other data sources to develop and disseminate data to farmers are in the early stages of testing and developing their farmer advisory services into products, as well as developing viable business models. Nonetheless, the region's experience in the precision agriculture use case suggests that drones can be a relatively low-cost solution for significantly increasing production volumes and can also deliver large environmental sustainability benefits through reduced water and fertilizer consumption.

Including in developed countries, drone technology has shown promising results in civilian infrastructure applications, with numerous successful feasibility studies and experiments performed to demonstrate the cost, time, and quality advantages versus traditional infrastructure inspection and monitoring methods. Many of these studies, as well as actual experience in Rwanda and Zanzibar, have found that drone technology can help deliver less time-consuming detection and inspection of power transmission and distribution infrastructure, road and bridge infrastructure inspection, and infrastructure construction management. However, cost and time savings may not materialize in all situations and will depend on the complexity and quality of the desired outcome and the regulatory context (for example, the ability to fly the inspection drones over live traffic), among other factors. Thus, in this use case, as in others, a more appropriate approach to analyzing the economic rationale for drones—and one that is more likely to be used by public and private sector decision-makers—would be *to select the approach that achieves the required level of service or quality at the lowest cost.* South Africa's experience also demonstrates that there are still obstacles to be overcome for local drone companies to readily engage with the power utility in providing inspection services over the long term.

Across the use cases, drone costs per kilometer or per other metric can be reduced, sometimes significantly, by improving the time utilization for each UAV; however, such efficiency will necessarily entail collaboration among different sectoral agencies or even actors managing supply chains in different sectors, such as health and agriculture. Another key consideration in the economics of future drone operations in the region relates to the need for regulatory support, or at least clarity, regarding the ability of UAVs to operate in the shared civilian air space, including beyond line of sight. In the absence of such clarity, UAV initiatives are bound to remain at the scale of donor-funded pilots, and investment in local technical capacity and necessary infrastructure—including infrastructure in ancillary services such as reliable internet and electricity connectivity services that enable UAV operations to run smoothly—will be discouraged.

Drone applications are rapidly evolving, and several use cases could grow in impact and scale over the coming years. In addition to the applications reviewed in this report, other less-well-documented use cases are emerging. Applications in disease monitoring and prevention have seen success in Ethiopia and Malawi. The COVID-19 pandemic has shown that countries with drone distribution networks in place, such as Ghana and Rwanda, could more easily scale up a drone-based pandemic response. In the field of conservation, UAVs hold much promise, particularly in monitoring illegal poaching and deforestation. Drones have the potential to transform business models, serve communities that are digitally online but physically isolated, and help society tackle some of the most pressing development challenges.

POLICY AND OPERATIONAL IMPLICATIONS

- Drones can solve critical issues in supply chains but need to be integrated as much as possible into existing systems rather than being developed as an entirely parallel system. Even if not directly integrated into existing public systems, donor-funded drone initiatives in the region should ensure technical capacity transfer to local stakeholders.

- The number and scale of UAV pilot projects must be significantly expanded to demonstrate a sustainable long-term business case.
- Cost-effectiveness of drone operations is critical in Africa; resources are scarce not only in public health systems but also in the private sector. The health pandemic of 2020 and 2021 and the resulting economic recession in many countries, with a projected decline in economic growth of between 2.1 percent and 5.1 percent in 2020 following positive growth of 2.4 percent in 2019 (Zeufack et al. 2020), mean that governments will have even fewer resources, rendering cost-effectiveness yet more critical.
- As suggested by examples in several countries in Sub-Saharan Africa, drone-based transport can effectively fill supply gaps in transportation services that arise in emergency situations such as COVID-19, for example, critical goods deliveries avoiding human interaction, aerial-image-based quarantine compliance monitoring, and others.
- Although drone initiatives in the region have mostly been donor funded or remain small scale, the private sector—drone manufacturers, operators, consultants, and commercial banks that can finance large UAS operations—is expected to play a much more prominent role in the future.
- Across the potential UAV use cases, the regulatory enabling environment—both support and predictability—is key. Regulations need to be harmonized across government entities in each country and across countries in the region; drone companies need to work closely with civil aviation authorities. The COVID-19 health pandemic has demonstrated that in most countries the enabling environment for quickly launching drones, including human resources with the right knowledge and skills to facilitate drone integration into the response, regulations, and air traffic management procedures, is not there yet.
- Developing the data ecosystem is important for both drone operators and public entities to understand what gaps drones are filling, what their impact on costs and societal outcomes may be, and what new UAV-based use cases might be viable as the context changes (for example, with the onset of widespread restrictions on in-person mobility and goods delivery systems).
- The World Bank can help governments define their national strategies for drone development. The African Drone Forum community could play a similar role at the regional level.

NOTES

1. A UAS consists of one or more UAVs, the associated equipment (launch and landing stations, batteries and charging equipment, flight control software, and so forth) as well as the software and personnel needed to control the UAV.
2. In the drones-as a-service model, drones are provided and operated for a flat monthly fee, which removes all other cost components.

REFERENCE

Zeufack, Albert G., Cesar Calderon, Gerard Kambou, Calvin Z. Djiofack, Megumi Kubota, Vijdan Korman, and Catalina Cantu Canales. 2020. *Assessing the Economic Impact of COVID-19 and Policy Responses in Sub-Saharan Africa. Africa's Pulse. An Analysis of Issues Shaping Africa's Economic Future*. Volume 21, April 2020. Washington, DC: World Bank.

Abbreviations

AIDS	acquired immune deficiency syndrome
ARV	antiretroviral
cm	centimeter
COSTECH	Tanzania Commission for Science and Technology
DALY	disability-adjusted life year
DBS	dried blood spot
DHS	Demographic and Health Survey
DrOTS	Drone Observed Therapy System
EID	early infant diagnosis
FH	flight hour
GCP	ground control point
GHSC-PSM	Global Health Supply Chain Program–Procurement and Supply Management
GPS	global positioning system
GSD	ground sample distance
HGSF	Home-Grown School Feeding
HIV	human immunodeficiency virus
IMF	International Monetary Fund
IV	intravenous
JSI	John Snow, Inc.
kg	kilogram
km	kilometer
km^2	square kilometers
LiDAR	light detection and ranging
MBTS	Malawi Blood Transfusion Service
MSD	Medical Stores Department
mt	metric ton
NBTS	National Blood Transfusion Services
PPH	postpartum hemorrhage
TB	tuberculosis
UAS	unmanned aircraft system
UAV	unmanned aerial vehicle
UNICEF	United Nations Children's Fund

xxii | UNLOCKING THE LOWER SKIES

VGI	volunteered geographic information
VL	viral load
VTOL	vertical takeoff and landing
WFP	World Food Programme
WHO	World Health Organization

All dollar amounts are US dollars unless otherwise indicated.

Introduction

CONTEXT: AFRICA'S TRANSPORT INFRASTRUCTURE NEEDS

Sub-Saharan Africa's transport infrastructure needs are enormous: even to sustain its current level of development, the region needs to spend $38 billion more each year on transport infrastructure, plus an additional $37 billion on operations and maintenance. A significant financing deficit separates Africa's current reality from the mobility that it both needs and aspires to. Most countries in Sub-Saharan Africa spend about 2 percent of gross domestic product (GDP) on all road network needs, and in general spend much more on capital investments than maintenance; about a quarter of the region's countries are not spending enough on road maintenance to cover routine maintenance activity (Cervigni et al. 2017). If countries spent 1 percent of their GDP annually on upgrading rural roads, even under optimistic GDP growth assumptions the rural access index—or the share of the rural population living within 2 kilometers of a road usable all year round—would only increase from 29 percent today to 46 percent by 2030 across Sub-Saharan Africa (Mikou et al. 2019). The rural access index is currently estimated at less than 30 percent in Burundi, Ethiopia, Malawi, and Tanzania, and between 50 percent and 60 percent in Uganda and Kenya (Iimi et al. 2016). Alternative solutions to ensuring connectivity, rural integration, and resilience of critical goods' supply chains are therefore needed in the short run, especially to serve the needs of remote communities.

Climate change is projected to bring about substantial changes in temperature and precipitation across Sub-Saharan Africa, with extensive effects on the existing road network. Climate change also poses a significant risk to Africa's bridges, across all projected future climates, and to the vital connectivity they provide for the transport network. Even assuming adequate maintenance regimes (thereby standardizing the analysis across countries), climate change will cause substantial disruptions in network connectivity and could lead to large increases in disruption time: in the worst climate scenarios, up to 2.5 times historic disruption time because of extreme temperatures, 76 percent higher because of precipitation, and 14 times higher because of flooding. In several countries, such as Mozambique and

South Sudan, even moderate changes in the climate will induce significant precipitation-related disruption (Cervigni et al. 2017). Many parts of Africa will face more intense precipitation, which can increase flooding frequency. These floods can overrun and erode roads, particularly unpaved ones (for example, Niang et al. 2014), cutting off entire communities from basic necessities such as food and medicine.

As demonstrated by the global economic and health crisis that began unfolding in the spring of 2020, essential goods supply chains across both Sub-Saharan Africa and many developed countries are highly vulnerable not only to climatic events but also to health pandemics. The COVID-19 (coronavirus) experience shows that traditional transport systems are easily disrupted by health concerns for transport and logistics sector workers and by government-imposed in-person travel restrictions. This vulnerability provides additional strong impetus for seeking complementary transport solutions that can fill the gaps.

THE POTENTIAL OF UNMANNED AERIAL VEHICLES

Considered an extension of the digital revolution, the proliferation of unmanned aerial vehicles (UAVs, or "drones") in nonmilitary applications is generating the technological capability to step in where traditional overland transport networks fail to perform. These airborne vehicles can increasingly be conceptualized and tailored to very specific mission parameters, take advantage of economies of scale from standardized building blocks, and be integrated within existing transport systems (Schwab 2016). Use cases that were either impossible or economically unsustainable for human-piloted aircraft are becoming technically possible and economically feasible (ITF 2018). Drones have the potential to counter supply-chain challenges in a range of sectors as costs continue to decrease. Cargo drones could augment existing rail, road, and sea systems; connect excluded communities; enhance the resilience of supply chains; and create new markets and services connecting urban and rural opportunities while also addressing the Sustainable Development Goals. But can this technology be effectively integrated into existing health and other goods supply chains and data-collection systems? Will it improve health outcomes and generate other public benefits? Can it be cost competitive with traditional transport technologies, especially those that benefit from explicit or implicit subsidies?

Initially emerging as tools for recreational purposes, since about 2015 drones have been increasingly used by firms across agriculture, inspection, construction, mining, and public safety, among others. More recently, technology companies have partnered with public agencies for goods delivery, leveraging semi-autonomous, beyond visual line of sight capabilities. Despite the vast opportunities for drone applications, the deployment of drones today remains predominantly small scale and concentrated in a relatively small number of special use cases, with each application field expanding the overall drone track record and revealing how this technology can improve the efficiency of other sectors compared with established technologies and practices.

The global drone logistics and transportation market generated more than $24 million in revenue in 2018, a number that is expected to grow to $1.6 billion in 2027. The value of prospective drone applications for global infrastructure projects is estimated at $44.2 billion, and prospective drone industry applications globally are valued at about $127 billion, measured by cost of labor and services that have high potential for replacement by drones (PwC 2017). In Europe, Single European Sky ATM Research reports that the drone market could reach $11 billion by 2035, of which more than half could be from delivery-related services, agriculture, and energy (SESAR 2016). However, as much as 9 percent of global drone shipments in 2022 are projected to be to Africa and the Middle East. In fact, it is in the world's South, notably Sub-Saharan Africa, where most progress in the civilian—in particular, medical—applications of drones has been made, an example of leapfrogging taking place on the continent similar to the adoption of mobile banking. If necessity is the mother of invention, rarely has it been more so than in the case of medical drone technology in Sub-Saharan Africa (McCall 2019).

Key industries using drone technologies include infrastructure, delivery and e-commerce, agriculture, mobility and logistics, energy, public safety and security, entertainment, insurance, mining and construction, and telecommunications. The predicted global value of drones by industry in 2021, or the value of business services and labor, is projected to be the highest in infrastructure applications ($45.2 billion), agriculture ($32.4 billion), and transport ($13 billion) (Chicourrat 2018). The use of drones in humanitarian action is a rapidly emerging field, with the scope of activities ranging from mapping, monitoring, and damage assessments to delivering essential items to remote or otherwise inaccessible locations (Soesilo et al. 2016).

USE CASES FOR DRONES IN EAST AFRICA

This report explores the economic and the nonmonetary rationale for introducing drone technologies in a select number of use cases: medical goods deliveries, food aid delivery, land mapping and risk assessment, agriculture, and infrastructure inspection. The first two correspond to cases in which physical deliveries of goods are performed using UAVs, and the latter three focus on the role of UAVs mostly as a technology for image and information gathering. Across these applications, the case for using unmanned aircraft systems is examined within the context of the so-called logistics objectives—total operating costs, speed, availability, and flexibility—as well as the human, or societal, objectives, which may include increased access to health care, lives saved, and increased environmental sustainability, among others. The report draws mostly on the experiences of East African countries, some of which have been at the forefront of drone technology implementation; additional insights from countries in other Sub-Saharan Africa subregions are featured where available.

This analysis is motivated by the need for a deeper and more systematic understanding of where and when UAV deliveries are most justified on economic and other grounds, such as for their environmental benefits or their potential for local capacity building. In turn, this increased understanding can be a step toward identifying viable long-term business models for UAV operations, their integration into existing systems, and their scaling up to be able to make a difference in meeting East Africa's many development needs.

In this report, the term "drone" encompasses all flying vehicles without a human operator on board, including vehicles with different levels of automation. These levels range from drones remotely piloted from the ground (within or beyond the visual line of sight) to fully autonomous drones requiring no human intervention in carrying out a mission. Other, more technical terms are also used to discuss drones in many countries and in aviation, such as remotely piloted aircraft systems and unmanned aircraft systems, in addition to UAVs.

REFERENCES

Cervigni, R., A. Losos, P. Chinowsky, and J. E. Neumann, eds. 2017. *Enhancing the Climate Resilience of Africa's Infrastructure: The Roads and Bridges Sector*. Washington, DC: World Bank.

Chicourrat, R. 2018. "Building a Safe & Sustainable Drone Ecosystem." AIRMAP, November 1. http://www.africandroneforum.org/wp-content/uploads/2020/01/robert-chicourrat -airmap.pdf.

Iimi, A., A. K. F. Ahmed, E. C. Anderson, A. S. Diehl, L. Maiyo, T. Peralta Quiros, and K. S. Rao. 2016. "New Rural Access Index: Main Determinants and Correlation to Poverty." Policy Research Working Paper 7876, World Bank, Washington, DC.

ITF (International Transport Forum). 2018. "(Un)certain Skies? Drones in the World of Tomorrow." International Transport Forum, OECD, Paris.

McCall, B. 2019. "Sub-Saharan Africa Leads the Way in Medical Drones." *The Lancet* 393 (10166): 17–18.

Mikou, M., J. Rozenberg, E. Koks, C. Fox, and T. Peralta Quiros. 2019. "Assessing Rural Accessibility and Rural Roads Investment Needs Using Open Source Data." Policy Research Working Paper 8746, World Bank, Washington, DC.

Niang, I., O. C. Ruppel, M. Abdrabo, A. Essel, C. Lennard, J. Padgham, and P. Urquhart. 2014. "Africa." In *Climate Change 2014: Impacts, Adaptation and Vulnerability*, edited by C. B. Field, V. R. Barros, D. J. Dokken, K. J. Mach, M. D. Mastrandea, T. E. Bilir, M. Chatterjee, et al., 1199–265. Cambridge, U.K., and New York: Cambridge University Press.

PwC (PricewaterhouseCoopers). 2017. "Clarity from above: PwC Global Report on the Commercial Applications of Drone Technology." PricewaterhouseCoopers, Warsaw.

Schwab, K. 2016. *The Fourth Industrial Revolution*. Cologne: World Economic Forum.

SESAR (Single European Sky ATM Research). 2016. *European Drones Outlook Study: Unlocking the Value for Europe*. Brussels: Single European Sky ATM Research.

Soesilo, D., P. Meier, A. Lessard-Fontaine, J. Du Plessis, and C. Stuhlberger. 2016. "Drones in Humanitarian Action: A Guide to the Use of Airborne Systems in Humanitarian Crises." FSD (Swiss Foundation for Mine Action), Geneva.

1 Medical Goods Deliveries

THE BIG PICTURE: THE DEMAND FOR MEDICAL GOODS DELIVERIES

Across East Africa, the need for lifesaving blood and medical goods deliveries is continuously rising as a result of the rapid population growth in most countries. Diseases that could be prevented with available medicines cause a large share of deaths of children and other vulnerable population groups. For example, about 75 percent of all child deaths in South Sudan are due to preventable diseases such as malaria and pneumonia; 1 in 10 children in the country dies before reaching his or her fifth birthday. Although Ethiopia has made significant progress in reducing maternal and under-five mortality and combating malaria and other diseases, maternal and newborn mortality remain unacceptably high. Every day, 480 children die of easily preventable diseases, and 353 women die in childbirth per 100,000 live births. In Uganda, 1 in 11 children dies before age five, compared with 1 in 7 in 2001. This progress notwithstanding, mothers, so critical to the survival of children, are still being lost to preventable conditions,[1] and maternal mortality accounts for nearly 12 percent of all deaths among women of reproductive age. Tuberculosis (TB) and malaria accounted for 6.4 percent and 3.9 percent of all deaths, respectively, in East Africa[2] in 2017,[3] with children, defined as those younger than age 15, accounting for more than 10 percent of all TB cases (WHO 2018).

Immunization rates in the region still vary widely, even in higher-income countries such as Kenya (figure 1.1), where in a few counties fewer than one-third of children under age one are fully immunized, according to the United Nations Children's Fund (UNICEF), the organization providing vaccine coverage for half of the world's children. Immunization coverage against *Haemophilus influenzae*, which causes illnesses such as pneumonia and meningitis, is particularly low in Somalia and South Sudan, where fewer than half of all one-year-olds are vaccinated. Immunization rates against measles are less than 60 percent in Kenya and Mozambique, while comprehensive coverage data are missing altogether for many of the region's countries according to the World Health Organization (WHO). In Uganda, there is significant unmet need for vaccinations in specific regions such as North Buganda, Busoga, East Central, and Islands, where fewer than half of all children have received all basic vaccinations according to 2016 Demographic and Health Survey (DHS) data. In South Sudan, vaccination rates are even lower according to UNICEF; for example, only 22 percent of children have received the measles vaccine. Globally, approximately 20 million children do not receive vaccines because they live in remote areas (UPDWG 2019).

The East Africa region also continues to battle the HIV/AIDS epidemic, with large associated needs for laboratory sample testing and regular supplies of antiviral medicine (map 1.1). Across East Africa, HIV/AIDS accounted for approximately 10 percent of all deaths in 2017, the second most important cause

FIGURE 1.1

Hib3 and MCV2 immunization coverage, 2018

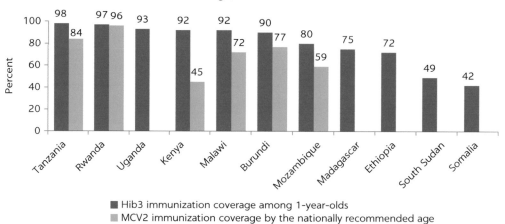

Percent
100
80
60
40
20
0
98
84
97
96
93
92
45
92
72
90
77
80
59
75
72
49
42
Tanzania
Rwanda
Uganda
Kenya
Malawi
Burundi
Mozambique
Madagascar
Ethiopia
South Sudan
Somalia
Hib3 immunization coverage among 1-year-olds
MCV2 immunization coverage by the nationally recommended age

Source: Data from World Health Organization's Global Health Observatory.
Note: Hib3 = *Haemophilus influenzae* type B; MCV2 = measles-containing-vaccine second dose.

MAP 1.1

HIV prevalence and treatment

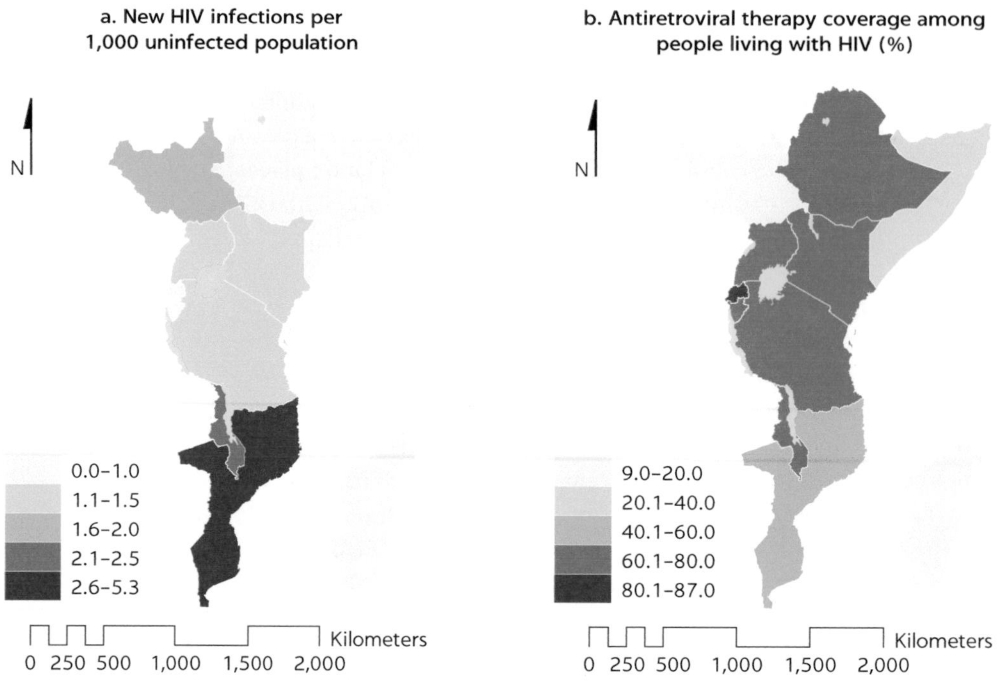

a. New HIV infections per 1,000 uninfected population
N
0.0–1.0
1.1–1.5
1.6–2.0
2.1–2.5
2.6–5.3
Kilometers
0 250 500 1,000 1,500 2,000
b. Antiretroviral therapy coverage among people living with HIV (%)
N
9.0–20.0
20.1–40.0
40.1–60.0
60.1–80.0
80.1–87.0
Kilometers
0 250 500 1,000 1,500 2,000

Source: Data from World Health Organization.

of death.[4] In individual countries, the importance of HIV in overall mortality is yet higher, reaching 17 percent in Kenya and Malawi and more than 24 percent in Mozambique. HIV prevalence in most East African countries significantly overlaps with TB prevalence, with 20–50 percent of new and relapse TB case patients also being HIV positive (WHO 2018).

Although some countries, such as Rwanda,[5] have made impressive progress in responding to the HIV epidemic, HIV testing in remote areas of most countries in East Africa remains difficult because of inefficiencies in blood sample delivery logistics, making it harder for large populations to start treatment programs in a timely manner. DHS data for Kenya for 2014 indicate that in individual regions only about half of all women have been tested for HIV and received results, with the figure as low as 37 percent in Mandera (and only 4 percent among men). DHS data for Uganda for 2016 illustrate that about a fifth of all women have never been tested for HIV in Bugiso, Bukedi, and Bunyoro; among men HIV testing coverage is even less complete, with more than 40 percent having never been tested in Bukedi and Karamoja. Only about 60 percent of all pregnant women in Busoga and East Central districts are tested for HIV during pregnancy and receive results and counselling. In the country overall, the Global Burden of Disease project estimates HIV/AIDS to be by far the single most important cause of death among children in the 5–14 age group, accounting for 42 percent of all deaths. The need for HIV testing and treatment, as well as for blood transfusion for postpartum hemorrhage (PPH), in Uganda is particularly high because of the country's consistently high fertility rate, reaching six children per woman in several regions.

This chapter discusses the potential and actual use of drones in medical goods transport in three geographic settings in the East Africa region—the Ukerewe District of Tanzania, Malawi, and Rwanda. Although their medical needs are not closely comparable, these settings share a challenging geography and topography for the transport of goods by traditional, land-based transport modes and therefore are good candidates for the introduction of alternative modes, such as drones, that can help overcome those challenges.

Ukerewe District, Tanzania

The health supply chain in the Ukerewe District of the Lake Victoria region of Tanzania is representative of the East Africa region when it comes to challenges at the intersection of demand for distribution of health products and poor rural accessibility. In Tanzania, largely preventable and treatable diseases such as malaria cause the death of 270 children younger than age five every day, and the under-five mortality rates range from 56 per 1,000 live births in the northern regions and Zanzibar to 88 per 1,000 live births in the lake regions such as Mwanza. And, although a 73 percent reduction in the maternal mortality ratio was recorded between 1990 and 2012 (870 and 232 deaths per 100,000 live births, respectively), previous gains are being eroded as the maternal mortality ratio increased to 556 per 100,000 live births in 2015.[6]

Tanzania continues to forge ahead with its vision to end AIDS by 2030. Access to lifesaving HIV prevention, treatment, care, and support for children and their families has been improved; the vast majority of pregnant and breastfeeding women with HIV have access to antiretroviral (ARV) treatment, and the share of infants diagnosed with HIV declined from 12.0 to 4.8 percent between 2011 and 2017. Despite this remarkable progress, Tanzania continues to carry 5 percent of the overall global burden of HIV among adolescents. Girls are disproportionately affected (UNICEF Tanzania 2018).

The shores of the Lake Victoria basin (map 1.2) are home to more than 35 million people. The Mwanza region is in the extreme north of mainland Tanzania, on the southern shore of Lake Victoria, and includes more than 86 inhabited islands in Lake Victoria. The population of Mwanza is approximately 2.8 million, which is second only to the commercial capital, Dar es Salaam, making it an area with one of the highest rural population densities in the world. Mwanza has a vast medical supply network, with 286 health facilities.

Vaccination rates in Mwanza in 2015 were at 89 percent for the Bacillus Calmette–Guérin vaccine administered to protect against TB, 87 percent for DPT3 (against diphtheria, whooping cough, and tetanus), and 88 percent for the measles vaccine. Only 70 percent of all children had received all basic vaccinations, compared with 86 percent in Dar es Salaam, according to the 2015–16 Tanzania DHS. According to the Tanzania HIV Impact Survey 2016–2017, the prevalence of HIV in the Mwanza region (where Ukerewe District is located) is 7.2 percent, significantly higher than the national average of 5 percent and the fourth highest in the country behind Iringa, Mbeya, and Njombe. Viral load suppression is only at 49.6 percent compared with the 90 percent goal for 2020 stated in the UNAIDS 90-90-90 strategy.[7] Only 63 percent of women and 43 percent of men in Mwanza have ever been tested for HIV and received results, compared with 72 percent and 57 percent, respectively, in Dar es Salaam, according to the Tanzania 2011–12 HIV/AIDS and Malaria Indicator Survey (TACAIDS, ZAC, NBS, OCGS, and ICF International 2013).

MAP 1.2

Population density in the Lake Victoria region

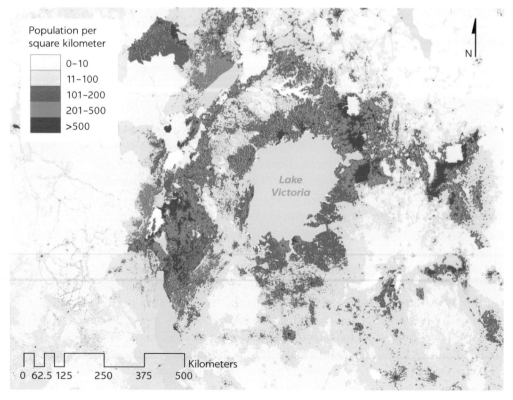

Source: WorldPop population data, 2020.

Ukerewe District, located partially across the islands of Lake Victoria, is one of the eight administrative districts of the Mwanza region. It consists of 38 islands, 34 of which are inhabited; 15 have permanent settlements, and 6 have health facilities. It has a total population of about 412,600, based on the United Republic of Tanzania 2012 Population and Housing Census projections for 2018 (Tanzania NBS 2013). Ukerewe is among the 81 priority councils for HIV interventions. Its health commodity needs have been estimated at 1.2 billion Tanzanian shillings (equivalent to approximately half a million dollars) per year (Ukerewe District Council 2017).

Ukerewe District has 37 health facilities, including one district hospital (in Nansio, on the main island), four health centers, and 32 dispensaries (map 1.3). The density of facilities is about 200 per 10,000 square kilometers (km²), with about 10,800 people served by each, on average. As part of the current study, field research was conducted in Ukerewe by John Snow, Inc. (JSI) over the fall and winter of 2019–20. Data on the consumption of various medical goods were extracted from the electronic logistics management information systems (eLMIS), along with data on the products out of stock at the health facility level. Analysis of the eLMIS was complemented by in-person visits to the health facilities in Ukerewe.

Demand for vaccines in Ukerewe is predictable; supply is usually based on annual forecasts divided into monthly allocations. The 2019 data show that the average volume delivered per facility (including Nansio District Hospital and the other health facilities of various sizes) was approximately 4,200 liters (for an overall total of 155,395 liters); however, the smaller facilities require much less than this average, about 70–100 liters annually. In addition, there are periodic

MAP 1.3

Ukerewe health facilities

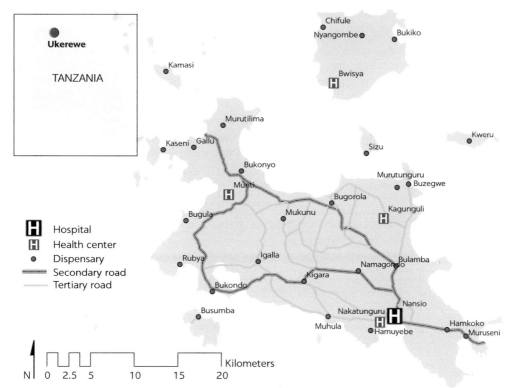

Source: Field research in Ukerewe District, Tanzania, fall/winter 2019–20.

campaigns—supplementary immunization activities—that focus on specific disease prevention, such as measles or polio, that are more focused and intense.

The second-largest demand volume in Ukerewe was for laboratory samples, with 32,131 liters delivered. The demand for emergency items, including oxytocin, magnesium sulfate, artesunate, intravenous (IV) fluids,[8] rabies vaccine, and antivenom, was equivalent to 10,788 liters by volume; IV fluids made up 62 percent of the overall volume. The demand for laboratory samples and emergency items, respectively, was approximately 78 liters and 26 liters per 1,000 inhabitants in Ukerewe. For blood samples that need to be tested to perform blood transfusions, 30–50 samples are transported each week from the Ukerewe health facilities to the National Blood Transfusion Services (NBTS) lab in Mwanza (on the mainland); during quarterly donation campaigns, demand exceeds 150 samples per week. In 2019, Ukerewe health facilities overall required the delivery of 943 liters of blood (samples and whole blood for transfusion), equivalent to 2.28 liters per 1,000 inhabitants.

Malawi

Malawi is representative of hard-to-reach rural areas with high demand for health assistance. In Malawi, maternal hemorrhaging accounts for up to 35 percent of all maternal deaths (Matemba 2019), and newborn deaths are currently the largest contributor to overall under-five deaths. Many of these deaths could be prevented with timely identification and treatment of infections and improved immunization coverage, especially among children who come from families living in poverty and rural areas. In some districts of Malawi, only half of all children have received all basic vaccinations, according to the 2015–16 DHS (Malawi National Statistical Office 2017).

Similarly to Ukerewe District, there are regions in Malawi with low coverage of HIV testing relative to its prevalence. The country's HIV prevalence ranges from about 5–6 percent in the Northern and Central Regions to in excess of 12 percent in the Southern Region. In 2014, approximately 10,000 children in Malawi died from HIV-related diseases, and fewer than half of all children who needed treatment were on it in 2018 (UNICEF 2018). HIV testing coverage is incomplete, and in many districts fewer than half of all women receive an HIV test and test results in a given year, and fewer than 80 percent have ever been tested for HIV (Malawi National Statistical Office 2017). As a result, nearly 40,000 children each year are born to HIV-positive mothers.

A study commissioned by UNICEF Malawi in 2018 (JSI 2018c) estimates the weekly demand for certain emergency and long-tail medical goods in the Nkhata Bay district in the Northern Region of Malawi (map 1.4). The estimated weekly average demand for viral load (VL), early infant diagnosis (EID), and TB specimen and result deliveries in the 22 surveyed facilities ranged from just 1 unit in the Bula, Chisala, Khondowe, and Nthungwa health centers in very low-population density areas to 13–15 in the Chintheche, Likoma Island, and Nkhata Bay hospitals. The weekly estimated shipment volumes of emergency and long-tail products ranged from 12 cubic centimeters (cm³) (a very small fraction of a kilogram [kg]) for the Nthungwa health center to 1,768 cm³ (0.5 kg) for Chintheche Rural Hospital. Finally, whereas the estimated weekly shipment of vaccines was less than 5,000 cm³ (about 0.5 kg) in several smaller health centers, it exceeded 20,000 cm³ (about 8.5 kg)

MAP 1.4

Nkhata Bay district health facilities surveyed in the 2018 UNICEF study in Malawi

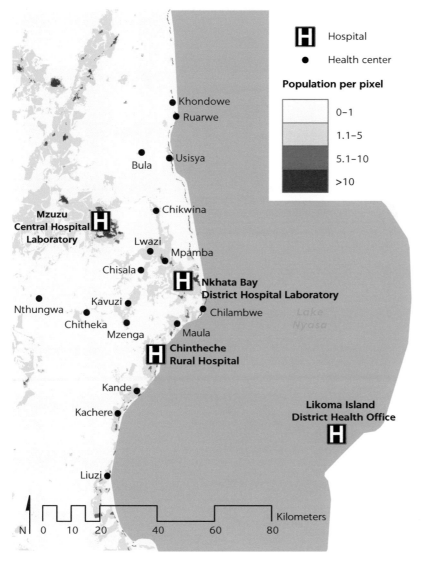

Sources: WorldPop population data, 2020; data from JSI (2018c).

for Chintheche Rural Hospital and 26,000 cm³ (about 11 kg) for Nkhata Bay District Hospital. For dried blood spot (DBS) and TB specimens and test results, peak demand was estimated at 131 specimens or results per week for all facilities combined.

More recent data received for the current study from Malawi's Ministry of Health and Population for six district hospitals and one community hospital (shown in map 1.5) present a complementary snapshot of the demand for emergency medical goods. The hospitals are located in less densely populated areas of the country's Northern (3 hospitals), Central (1 hospital), and Southern (3 hospital) Regions and have a direct catchment area population of approximately 307,000. All the hospitals offer surgical, neonatal, and obstetric care services; administer vaccinations; and perform laboratory diagnostics in addition to providing outpatient consultations.

MAP 1.5

Hospitals covered by July–September 2019 data received from Malawi Ministry of Health and Population

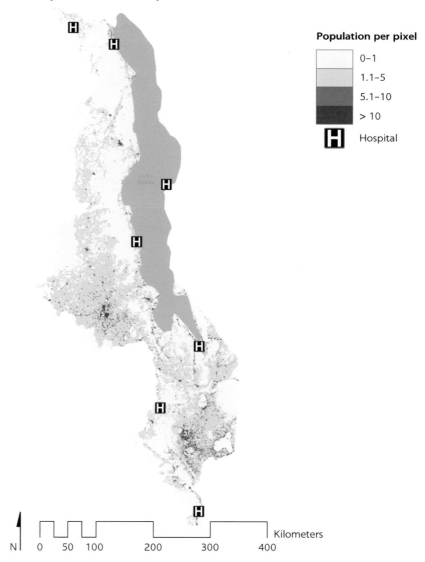

Source: WorldPop population data, 2020.

Between July and September 2019, the hospitals recorded 182 cases of PPH; of these, 135 required blood transfusions and 81 cases required transfusions of three or more units of blood or other blood products (platelets, cryoprecipitate). On average, each PPH case requires 2.7 units of blood for total blood demand of more than 490 units across the seven facilities. Over the same time, the facilities had 152 cases of neonatal sepsis due to factors such as prolonged labor and typhoidal *Salmonella*. Among other emergency-type situations, the facilities recorded 37 cases requiring the administration of antivenom treatment; 148 cases of stroke (ranging between 2 and 84); and 1,070 cases requiring the administration of anti-rabies vaccination or intravenous immunoglobulin (ranging from none to 354). In addition to urgent medical needs, the surveyed hospitals during the three months treated 36,894 cases of malaria (reaching 9,969 cases in the heaviest-burden facility) and 886 cases of TB (ranging from none to 700 per facility).

PRESENT COSTS AND MODALITIES

Ukerewe District, Tanzania

Medical goods distribution in Ukerewe is subject to substantial logistical and transportation challenges. The specifics of the district's geography (only 70 percent of the total area is accessible by land), the absence of a high-quality road network, and the reliance on local boats for transport between the islands are some of the key drivers. The infrastructure deficit means that, despite best efforts, delivery coverage can never be quite complete: approximately 20 percent of the health facilities may be inaccessible by land transport for an average of 10 percent of the time in a given year. The result is lost lives and reduced ability to mitigate disease outbreaks.

As part of the facility-level analysis in Mwanza conducted for this study, the various flows of medical products were mapped, along with the associated transport costs and identification of the stakeholders or entities responsible for transporting the specific goods or for covering the costs of transportation. The mapping of the individual goods flows is illustrated and explained in more detail in this chapter; however, as a general rule, the Medical Stores Department (MSD), a parastatal entity, is tasked with delivering essential medicines, medical supplies, and emergency medicines to all health facilities across Tanzania, with about 7,500 delivery points across the country. MSD consignments typically travel from the MSD head office in Dar es Salaam by road to the Zonal Store in Mwanza over approximately 1,200 km, followed by an overland trip to Nansio District Hospital in Ukerewe Island (250 km) (figure 1.2). The consignment is then off-loaded and loaded to an SUV and ferried through to the smaller island health facilities. To reach the facilities located in Ukara, it is necessary to take a ferry from Nansio, which makes only one trip per day (Rugambwa Bwanakunu 2018). All public sector health facilities in Mwanza receive supplies from the MSD in Mwanza city. The Bugando Medical Centre, also located in Mwanza city, is a referral, consultation, and university teaching hospital for the Lake and Western zones and serves a catchment population of more than 14 million people. Laboratory samples (including VL samples and DBS) that are collected by health facilities must be sent and tested at the center. Because of the geography

FIGURE 1.2

Schematic of transport supply chain for medical goods distribution to Ukerewe islands

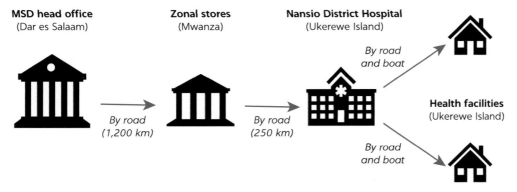

Source: Field research in Ukerewe islands, Tanzania, fall/winter 2019–20.
Note: MSD = Medical Stores Department.

of Ukerewe District, it is easier to send blood units to Ukerewe than to refer patients who need blood to Mwanza during emergencies.

Lifesaving medicines

All routine distribution goods—medicines that are routinely ordered through the Integrated Logistics System, including for a variation of TB—are managed by the MSD (figure 1.3). The MSD in Mwanza distributes Integrated Logistics System packages up to the last mile every other month; it also performs parallel deliveries for the TB System, Lab System, and ARVs to Nansio District Hospital every other month. Medicines for TB are picked up directly from Nansio District Hospital by the respective health facility staff. Efforts are increasing to decentralize multidrug-resistant TB treatment from a few centralized sites to many facilities because doing so has been shown to improve access and increase treatment adherence (Wright et al. 2018).

In addition to routine orders, lifesaving medicines are sometimes ordered on an emergency basis, which is associated with a slightly different transport supply chain. "Long-tail" products are those having small and unpredictable demand at individual health facilities. They share characteristics similar to those for serving rare blood types and include treatments for relatively unusual patients who are exposed to potentially fatal viruses or toxins, such as rabies postexposure prophylaxis, snake antivenom, and tetanus toxoid (for an emergency booster case). Long-tail products might also encompass treatments for severe conditions of more common diseases and are therefore unpredictable. They include medicines to treat multidrug-resistant TB, second- and third-line ARV drugs for HIV, artesunate injection for treating severe malaria, and furosemide injection for severe edema or hypertension. Because these conditions are unpredictable, in general, the products are not widely distributed or widely stocked at the health facility level.

FIGURE 1.3

Transport supply chain for routine distribution

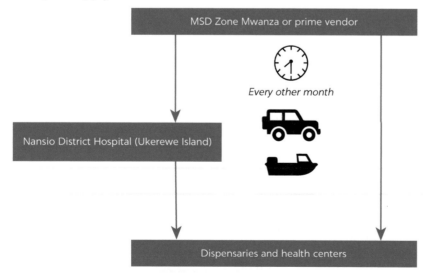

MSD Zone Mwanza or prime vendor

Every other month

Nansio District Hospital (Ukerewe Island)

Dispensaries and health centers

Who pays

- Facilities pay a delivery fee to MSD.[a]
- For vertical program[b] items (ARVs, malaria drugs, and others), donors pay MSD a distribution fee (percentage of the value of the medicines supplied); no transport cost in this case is incurred by health facilities.

Who does what

- MSD distributes directly to health facilities, except for TB medicines, which are picked up from Nansio District Hospital by health facility staff or distributed using the district medical office vehicle.

Source: Field research in Ukerewe District, Tanzania, fall/winter 2019–20.
Note: ARV = antiretroviral; MSD = Medical Stores Department; TB = tuberculosis.
a. The Ministry of Health, Community Development, Gender, Elderly and Children allocates funds to each facility. These funds are put in the respective facility's account at the MSD and cover the costs that the facility incurs when purchasing commodities from the MSD plus the delivery cost.
b. "Vertical programs" are typically developed for specific health conditions, whereas "horizontal systems" ensure general medical services.

The delivery of emergency orders, including any essential medicine requested outside of routine orders, is the responsibility of the MSD. As confirmed by the field research conducted for this study in the fall and winter of 2019–20, the commodity flow in this case is direct from the MSD to health facilities (figure 1.4).

Previous research in Mwanza has reported that there is no antidote for snakebites or anti-rabies vaccines in many of the health facilities on Lake Victoria islands, requiring that, in an emergency, the patient be transported to the mainland, which takes several hours. Ferries between the islands are infrequent, costly, and can be dangerous—a high-profile ferry disaster left more than 200 dead in September 2017—while speedboats, which take only half an hour to get to the mainland, are prohibitively expensive (Wakefield 2019). A 2018 study reported that long-tail products are often not stocked at the health facility level, which results in potentially high hidden demand (Wright et al. 2018).

Health facility–level research conducted for the current study in fall and winter 2019–20 suggests that lifesaving items such as magnesium sulfate and oxytocin are generally available in the visited health facilities, and there were no stock-outs in 2019, driven by the fact that any stock-outs in a given facility can be quickly fixed by redistributing the needed goods from other Ukerewe facilities. However, antivenom is only kept in stock at Nansio District Hospital, and all snakebite cases are referred to that facility. This medicine is expensive and demand is small; therefore, it is easier to keep supplies at the referral facility instead of distributing it to each facility and risking expiry.

In the field research undertaken for this study, several data types were analyzed to estimate the overall and per-trip costs of transporting the various lifesaving items in Ukerewe District. These data include information on two years of historical demand, by quarter; information on two years of stock-out rates, by quarter; historical caseload for selected patient emergencies; and information collected from health facility visits on current transport modes and detailed costs to distribute commodities between (1) Mwanza and Nansio District Hospital and health facilities and (2) between Nansio District Hospital and health facilities.

FIGURE 1.4

Transport supply chain for lifesaving items supplied on an emergency basis

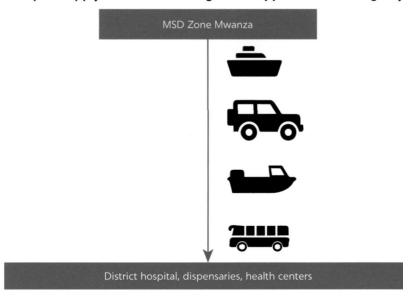

Who pays

- Facility staff and district staff: round-trip transport plus cargo fees.
- MSD covers costs to transport from the MSD Zone Mwanza to the ferry.
- Overall transport costs depend on the weight and volume, mode of transportation used, and number of days staff stay in Mwanza.

Who does what

- MSD transports from the warehouse to the ferry.
- Health facility staff travel to the ferry in Mwanza to collect the items, then travel back to the facilities.

MSD Zone Mwanza

District hospital, dispensaries, health centers

Source: Field research in Ukerewe District, Tanzania, fall/winter 2019–20.
Note: MSD = Medical Stores Department.

For lifesaving medicines, the estimated transport cost per trip varies between $1[9] for the eight Nansio ("big island") facilities located within 10 km of Nansio District Hospital to $74 for the facilities on the small islands (table 1.1). Although the total number of annual trips is by far the highest to the 18 health facilities located on the big island but farther than 10 km from Nansio, the overall annual transport cost is highest for the small island facilities, at $6,665. The total annual cost of delivering lifesaving medicines to all facilities in Ukerewe is estimated to be $11,152.

Vaccines

Vaccine supply chains in Ukerewe District work in parallel to the MSD system, with the Immunization and Vaccine Department being the primary manager of the system. The Regional Immunization and Vaccine Officer, who is a custodian of the Regional Vaccine Store, is responsible for supplying vaccines and related supplies to District Immunization and Vaccine Officers once every quarter. In turn, the District Immunization and Vaccine Officer is responsible for monthly delivery to health facilities (the last mile). The Regional Vaccine Store is located at the Regional Medical Office in the Sekou Toure regional hospital in Mwanza; the District Vaccine Store is located in Nansio District Hospital (figure 1.5).

Vaccines for routine immunization are typically delivered in cold boxes using a combination of boat and car transport. Last-mile transportation is the most challenging and accounts for the third-highest cost incurred by councils in Tanzania (JSI 2018a). Vaccine delivery usually takes about 10 days to complete. It requires cold chain maintenance throughout the supply chain, and, with the transportation challenges, cold chain maintenance is usually put at risk.

Cold chain equipment, management, and maintenance are major cost drivers in the vaccine supply chain (Wright et al. 2018). Older health facility refrigerators are prone to breaking down, and many rely on kerosene or liquefied gas for fuel, which is expensive to procure and transport. Cold boxes must be stocked at distribution sites, and freezers must be used to condition cold boxes in preparation for use and to freeze ice packs. Preparing cold boxes takes time, which affects both the lead time and total cost of distribution. Furthermore, a major challenge with many cold boxes currently in use is the risk of freezing vaccines, which can render many antigens ineffective. Closed-vial wastage also affects inventory holding costs. Data from Tanzania suggest it is not a significant problem (1–2 percent of total volume annually), but data on cold chain equipment functionality show as many as 10 percent of refrigerators to be nonfunctional at any given time. If a refrigerator at a health facility is

TABLE 1.1 Estimated number of trips and costs for lifesaving medicines, by destination, using current transport methods

DESTINATION	TOTAL TRIPS/YEAR	COST/TRIP ($)	ANNUAL COST ($)	VOLUME OF MEDICINE/ TRIP (LITERS)	ASSUMPTIONS
Big island (near)	80	1.30	104	1.5	• Includes oxytocin, magnesium sulfate, artesunate, and intravenous fluids sourced from Nansio District Hospital. No stock-outs at Nansio District Hospital for these items. • Rabies and antivenom are sourced from the Medical Stores Department or Prime Vendor and are *in addition* to the trips, costs, and volumes in this table.
Big island (far)	180	24.35	4,383	1.5	
Small islands	90	74.06	6,665	1.5	

Source: Field research in Ukerewe islands, Tanzania, fall/winter 2019–20.

FIGURE 1.5

Transport supply chain for vaccines

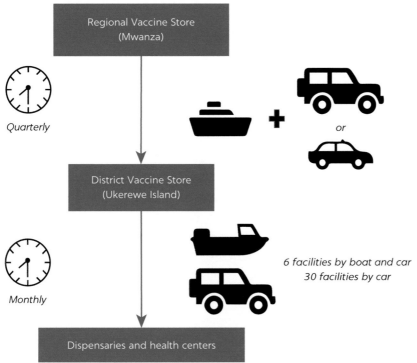

Who pays

- The cost of routine distribution from regional to district facilities is included in the Regional Medical Office annual budget.
- The cost of routine distribution from district to health facilities is budgeted in the Comprehensive Council Health Plan.

Who does what

- Routine distribution from regional to district facilities is done by Regional Immunization and Vaccine Officer; from district to health facilities is done by District Immunization and Vaccine Officer.
- In cases of vaccine shortages at the district level, the District Immunization and Vaccine Officer travels to the Regional Vaccine Store using public transport (ferry and taxi).
- For shortages at the health facility level, health facility staff travel to the District Vaccine Store with a cold box to pick up vaccines.

Source: Field research in Ukerewe District, Tanzania, fall/winter 2019–20.

nonfunctional, health care workers must travel to the district or neighboring health facilities to store vaccines.

Transportation costs in Ukerewe were assessed for all routine distribution vaccines; emergency-type vaccines were analyzed separately within the life-saving medicines category. Because of the low associated volumes, the transport cost per trip is by far the highest for vaccine deliveries to Nansio District Hospital, at $226 (table 1.2). However, given the much higher number of trips required in a given year, the total annual transportation cost is highest for deliveries to the big island, at $8,576. The total transportation cost for delivering vaccines to all health facilities in Ukerewe District in 2019 was an estimated $16,554.

Blood for transfusion

Previous research on medical goods supply chains in Mwanza analyzed blood for transfusion jointly with blood samples transported for diagnostic purposes. In the present study, the two are addressed separately. The supply chain for blood for transfusion for Ukerewe District is managed by the NBTS. In Ukerewe District, only four health facilities conduct blood transfusions: Nansio District Hospital, Bwisya Health Center, Muriti Health Center, and Kagunguli Health Center.

Safe blood for transfusions is often in scarce supply, especially outside central blood banks and for rare blood types (such as Rh negative) and supporting products (platelets, plasma). Blood for transfusion is also relatively difficult to store and has a limited shelf life. Hospitals may have various types of packed red blood cell products and different types of plasma available to prevent delays in emergency transfusions. Blood groups with relatively predictable demand may be

TABLE 1.2 Estimated number of trips and costs for vaccines, by destination, using current transport methods

DESTINATION	TOTAL TRIPS/ YEAR	COST/TRIP ($)	ANNUAL COST ($)	VOLUME/TRIP (LITERS)	ASSUMPTIONS
District hospital	12	226.42	2,717	38,849	• Quarterly distribution to Nansio District Hospital (1 distribution requires three trips) • Monthly distribution from Nansio District Hospital to all facilities using milk runs
Big island	312	27.49	8,576	349	
Small islands	108	48.71	5,261	349	

Source: Field research in Ukerewe District, Tanzania, fall/winter 2019–20.

stocked at the hospital- or health-center level if cold storage is available. However, because of the challenges in predicting demand for rare blood types, these items are even more difficult to stock at point of transfusion. Although the shelf lives of packed red blood cells (42 days) and plasma (1 year) are relatively long, other products, such as platelets (5 days) and thawed plasma (5 days), may be wasted when demand is low. Whole blood plasma must be frozen at −20 degrees Celsius within 24 hours of preparation, which presents its own transportation challenges. In a patient with severe bleeding, a massive transfusion (more than 10 units in 24 hours or 5 units in 60 minutes) may be needed, which can often rapidly deplete a hospital's blood supply (Thiels et al. 2015). These factors make it difficult to keep stock in sufficient quantities close to where it may be needed.

JSI (2018a) collected data in the fall of 2018 on the blood supply chain in Mwanza, studying the transport of samples from Ukerewe to NBTS in Mwanza for testing and emergency blood for transfusion (whole blood sent from NBTS in Mwanza to Ukerewe District). According to that research, blood for transfusion accounts for a relatively small share of overall costs; however, the lower overall financial costs are offset by the more emergency nature of this supply chain (timeliness is much more essential). Health facility–level research conducted in Ukerewe for the current study in the fall and winter of 2019–20 suggests that whole blood for transfusion sometimes does not incur any explicit or up-front cost because some transporters carry it for free, seeing this service as being part of their corporate social responsibility (figure 1.6). The data collected at the health facility level suggest that the big island receives by far the highest number of blood deliveries per year (192)—more than Nansio District Hospital and the small islands combined. However, the transportation cost per trip is for blood deliveries to Nansio District Hospital and the small islands (table 1.3), and the overall annual transportation cost is highest for deliveries to the small islands, at $7,496. Across Ukerewe District, the transport of blood to health facilities cost $17,555.

Laboratory samples

Diagnostic samples include, for example, DBS for EID or VL, whole blood or serum for HIV-related testing and monitoring, sputum for TB diagnostics, histology samples for biopsy testing, and other sample types for disease surveillance systems (Wright et al. 2018). Laboratory samples in Mwanza are managed by the Diagnostic Services Station.

Laboratory samples require good sample management to ensure integrity of results. Most false outcomes are due to sample integrity being jeopardized during the transportation process. The rugged island terrain around Lake Victoria means that samples must be collected and results delivered by ferry and motorbike. This process is slow and unstable, and samples may be exposed

FIGURE 1.6

Transport supply chain for blood for testing and whole blood for transfusion

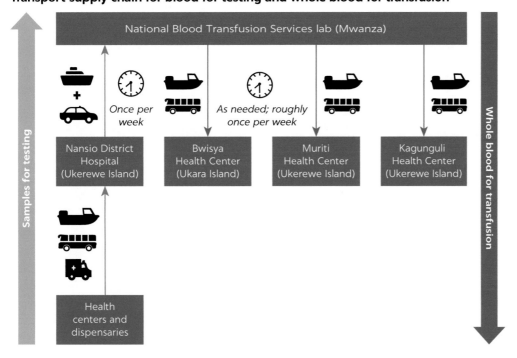

Who pays

- Samples for testing: Ukerewe District Council pays for transport.
- Whole blood for transfusion: National Blood Transfusion Services; some transporters carry it for free.

Who does what

- Samples for testing: Packaging and barcoding of samples done at Nansio District Hospital; sample transport is done by health workers.
- Whole blood for transfusion: Transported by either Mwanza health worker or transporter; depending on who performs the transport, it can be either a round trip or a one-way trip.

Source: Field research in Ukerewe District, Tanzania, fall/winter 2019–20.

TABLE 1.3 **Estimated number of trips and costs for blood samples and blood for transfusion, by destination, using current transport methods**

DESTINATION	TOTAL TRIPS/ YEAR	COST/TRIP ($)	ANNUAL COST ($)	VOLUME/TRIP (LITERS)	ASSUMPTIONS
District hospital	48	106.96	5,134	6.0	• Only three health centers and Nansio District Hospital carry out blood transfusion services.
Big island	192	25.65	4,925	4.2	• Nansio District Hospital transports samples
Small islands	96	78.08	7,496	9.6	only to Mwanza; no routine blood distribution from Mwanza (only for rare blood types).

Source: Field research in Ukerewe District, Tanzania, fall/winter 2019–20.

to rough transit and intense heat over several hours, possibly compromising sample quality. In Ukerewe, too, the major setback for VL samples is sample integrity, which is jeopardized during transportation using motorcycles. Compromised sample integrity results in false outcomes, complicating clinical management of persons with HIV.

Ukerewe District has organized sample collection in such a way that health facilities with no capacity to process samples ("spokes") ship them daily for initial processing at facilities with such capacity, called "hubs" (figure 1.7). Hubs have infrastructure such as refrigerators, cold chain, and centrifuges.

FIGURE 1.7

Transport supply chain for laboratory samples

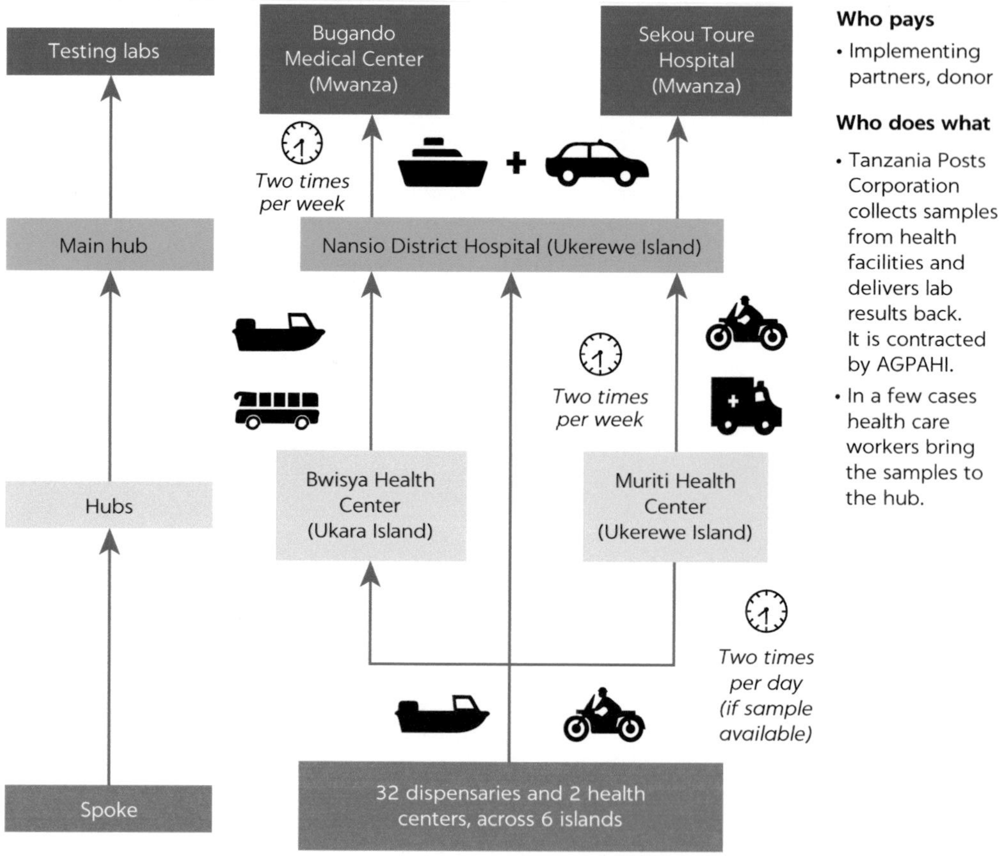

Source: Field research in Ukerewe District, Tanzania, fall/winter 2019–20.
Note: AGPAHI = Ariel Glaser Pediatric AIDS Healthcare Initiative, a Tanzanian nongovernmental organization.

Samples must be processed not more than five hours after collection and need to reach the testing lab no later than five days after collection. The Muriti and Bwisya hubs submit their samples to the main hub, Nansio District Hospital; the samples are then transported to Mwanza twice a week. Each hub collects an estimated 50–100 samples a week.

VL sample transportation is starting to be outsourced to private sector players: SkyNet is currently operating in Mwanza, and Tutume Worldwide Ltd in Kagera. In the Lake Zone, the active partners are Management and Development for Health and the Ariel Glaser Pediatric AIDS Healthcare Initiative (JSI 2018a).

In the current study, transportation costs in Ukerewe were estimated for VL testing and DBS strips. The far (more than 10 km from Nansio) health facilities on the big island require the largest number of annual trips in this goods category (2,592) and entail the highest overall transport cost ($63,110); however, the transport cost per trip is much higher for sample transport from Nansio District Hospital and the small island facilities (table 1.4). Lab sample pickup is the single largest transport cost category for all medical goods in Ukerewe, at $114,675 per year.

In summary, lifesaving medicines account for about 7 percent, or the smallest single share of any goods category, of overall transport costs associated with

TABLE 1.4 **Estimated number of trips and costs for laboratory samples, by destination, using current transport methods**

DESTINATION	TOTAL TRIPS/YEAR	COST/TRIP ($)	ANNUAL COST ($)	VOLUME/ TRIP (LITERS)	ASSUMPTIONS
District hospital	144	106.96	15,402	4.0	• District hospital: Three trips per week
Big island (near)	1,152	1.30	1,503	0.6	• Health facilities send samples one or more times per week
Big island (far)	2,592	24.35	63,110	0.6	• "Near" facilities are within a 10-km drive of Nansio District Hospital; no per diem, low delivery cost
Small islands	468	74.06	34,660	1.2	

Source: Field research in Ukerewe District, Tanzania, fall/winter 2019–20.

FIGURE 1.8

Share of individual medical goods and destinations in overall annual medical goods transport costs in Ukerewe

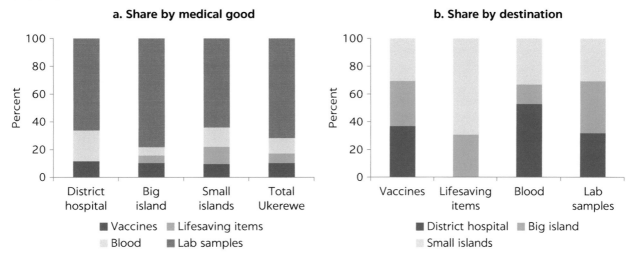

Source: Field research in Ukerewe District, Tanzania, fall/winter 2019–20.

transporting the studied medical goods to all health facilities in Ukerewe in 2019, although the respective share is significantly higher for the small islands, at 12.3 percent (figure 1.8). Blood deliveries account for the second-lowest transport cost in Ukerewe overall, at 11 percent; however, for Nansio District Hospital, this category accounts for as much as 22.1 percent of all transport costs. The cost of transporting vaccines makes up between 9.7 percent (small islands) and 11.7 percent (district hospital) of all transport costs. Finally, lab sample pickup is associated with by far the highest share of overall transport costs, at between 64.1 percent (small islands) and 78.2 percent (big island).

However, for individual medical goods categories, the different destination types have very different degrees of importance in the overall medical goods transport costs in Ukerewe. For example, although the cost of transportation to the small islands accounts for most of the overall transport cost associated with deliveries of lifesaving items, for blood the cost of transport to Nansio District Hospital makes up the largest share, partly because blood is transfused only at Nansio District Hospital and three other facilities.

Malawi

Quality care of HIV patients depends on early diagnosis, which requires taking DBS specimens from the health center to the central laboratory for testing.

In remote regions of Malawi, terrain, infrastructure, and resource limitations delay—and can even prevent—lifesaving diagnoses and medicine deliveries to hospitals and health centers. Generally, the district hospitals tend to be relatively well connected by road to the Central Medical Stores in the largest cities, but the health centers and health posts face significant transport challenges and experience frequent stock-outs (figure 1.9). Similar obstacles exist in countries across the region and the developing world more generally.

Malawi's current system for transporting laboratory samples for EID of HIV relies primarily on road transport. DBS specimens are collected from more than 700 health facilities and sent by ground and water transport to district hospitals; from there they are sent to one of nine laboratories (Malawi Ministry of Health 2016). It currently takes an average of 11 days to get samples from health centers to a testing lab (figure 1.10), and it can take up to two months for the results to be delivered back; sometimes samples get lost on the way. Median turnaround time for EID is 29 days, with 75 percent of results returned between 22 and 41 days after sample collection. For VL tests, median turnaround time is 37 days (Phillips et al. 2016). Factors such as the high cost and price variability of diesel fuel ($1.00–$1.50 per liter, with significant fluctuations), poor infrastructure conditions, inaccessibility of some areas because of flooded roads, and limited distribution schedules have resulted in extreme delays in lab sample transport, which

FIGURE 1.9

Schematic of medical goods delivery system in Malawi and associated challenges

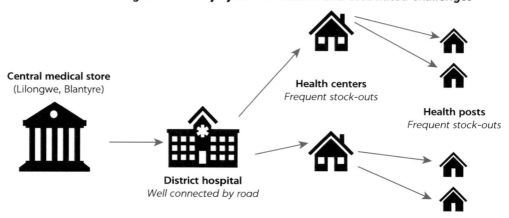

Source: Interviews with UNICEF Malawi staff.

FIGURE 1.10

Turnaround time for EID, VL, and TB specimen collection and delivery of test results: Illustrative example for Nkhata Bay, Malawi

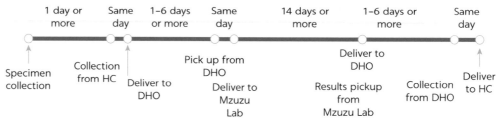

Source: Adapted from UNICEF (2018).
Note: DHO = district health office; EID = early infant diagnosis; HC = health center; TB = tuberculosis; VL = viral load.

is a significant impediment to scaling up pediatric ARV therapy effectiveness (UNICEF 2016).

In northern Malawi, most mainland facilities are served directly by motorcycle courier twice per week on a fixed schedule. One facility and the reference laboratory at the Mzuzu Central Hospital are served once per week from the Nkhata Bay District Health Office. The two inaccessible mainland facilities are served indirectly, transporting their items by ferry or on foot to another facility that acts as a transfer point; the Likoma Island facilities use the local ferry to reach the mainland. Several other private facilities manage their own specimen and result transport to the district health office (JSI 2018c). Lab samples from the islands in Lake Malawi may wait at St. Peter's Hospital for up to two weeks before being picked up and transported by boat to Nkhata Bay District Hospital on the mainland for processing. The results could then take up to eight weeks to return to the islands, delaying diagnosis and treatment. Even then, nearly half the results never make it back because the samples or the results are lost during transport; moreover, the amount of elapsed time may render patient monitoring largely ineffective.

In Kasungu District in central Malawi, six health facilities serve not only the district's own population of about half a million but also those of nearby districts, where stock-outs of various medical products are common. The district becomes physically isolated for three to five months each year because of weather conditions.

Data for the third quarter of 2019 from the Ministry of Health and Population for seven hospitals across Malawi illustrate the current time requirements and delays in administering various lifesaving medicines, even though these hospitals tend to be better connected and supplied than the average health center or health post. Each of the hospitals over the period July–September 2019 had an average of 4.7 patients with PPH that received blood transfusions more than an hour after the PPH diagnosis. At some of the facilities, timely administration of blood transfusion was hampered not only by the lack of readily available blood products but also by the limited availability of needles, IV lines, and personnel capable of administering IVs. For neonatal sepsis, five cases across the seven hospitals involved a delay in receiving antibiotics or had no antibiotics available. Among snakebite cases, the average time recorded from the patient's presentation at hospital and administration of antivenom was as long as 24 hours in the worst cases. Finally, for rabies treatment, although most of the surveyed hospitals experienced no delays in administering anti-rabies vaccination, one of the facilities reported 50 cases in which patients did not receive their rabies vaccination or their rabies vaccination was delayed because of stock-outs.

Rwanda

As a "land of a thousand hills," Rwanda faces challenges in ensuring that health commodities are made available when there is a critical need and time is of the essence (McCall 2019). Before the introduction of drone-based blood deliveries in Rwanda in 2016, and for those facilities that still remain unserved by the drone delivery system, transportation of blood to health facilities could take up to five hours by car or motorbike, often requiring staff such as lab technicians to leave their work stations to follow up on the order and delivery. The land-based transport system is very unreliable, especially during the rainy season; roads can become impassable, preventing trucks, vans, and even motorbikes from successfully completing their deliveries. The few blood collection and distribution centers in Rwanda are mostly located around referral hospitals in major cities. Most health facilities are not adequately equipped to store medical products that require careful handling, transport, and storage, and blood supplies are replenished about twice a week (USAID 2017).

INTEGRATING DRONES INTO EXISTING SUPPLY CHAINS

In what supply chains and route types can unmanned aerial vehicles add value?

For drones to augment the road network in Sub-Saharan Africa, they need to be integrated into the existing medical supply chains and distribution channels rather than operate in isolation or in parallel. Sub-Saharan Africa's annual road network investment needs amount to tens of billions of dollars, considering the current low shares of rural populations served by all-season roads. Drones can augment this incomplete road network. However, to add value, drones need to operate where supplies are stored and samples are collected, as well as where they are used—or where samples are received and test results generated (Aryal and Dubin 2019). The introduction of unmanned aerial vehicles (UAVs, or "drones") also assumes the existence and use of funding either for an insourced or outsourced model. For an insourced model, funding is needed to purchase and maintain UAVs to keep them flightworthy and for engineers to operate and maintain them (ISG 2017). Historically, it has been difficult to integrate storage and land transport across different health programs, so there is no reason that it will be any easier for unmanned aircraft systems (UASs)[10] (Wright et al. 2018). However, there are numerous examples of well-established transport regimes being ousted by disruptive inventions that make tremendous efficiency gains in contrast to the established one, for example, the transition from nonmotorized transport (by foot and carriages) to the railway and tram system. Creative use finds more and more deployment fields, and, with time, the new dominant technology creates its own demand. For example, the demand for leisure and holiday trips or logistics services evolved as a "car-technology-specific demand" or "truck-technology-specific demand" (Müller, Rudolph, and Janke 2019).

Given the major challenges that many public health supply chains face—high transport costs, chronic stock-out rates coupled with notable waste, and inefficiency caused by fragile or fragmented supply chains—UASs may offer a reliable last-mile delivery system for selected scenarios or use cases (USAID 2017), for example, to the many remote areas in the East Africa region that may be difficult to access for any number of reasons (Dirks 2017). Such reasons may include the following, among others:

- *For ground transportation*—lack of roads, bad road conditions, insufficient local availability of vehicles, rainy season, refusal of access by the military or the government, and security issues. For example, in Malawi, flooding severely affected thousands of people in 2015 in the southern part of the country (map 1.6). In 2019, more than 800,000 Malawians living in the south of the country were affected by massive floods, according to UNICEF Malawi.

MAP 1.6

Flood-affected population in Malawi, March 2015

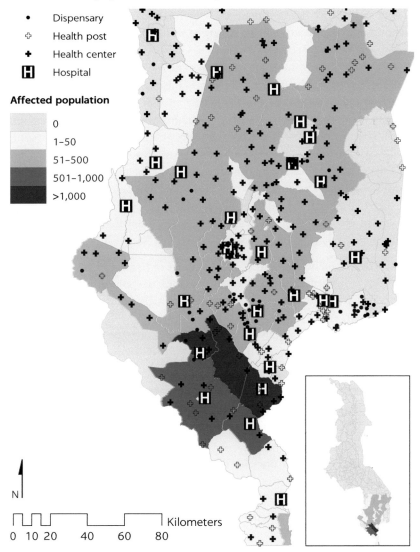

Source: Based on Netherlands Red Cross data.

- *For air transportation*—lack of landing zones, restrictions, and weather conditions (for example, fog).
- *For all traditional modes*—severe restrictions on in-person travel and deliveries due to health risks and government restrictions in the context of major health pandemics such as COVID-19 (coronavirus).

Unlike improvements in overland infrastructure or health financing that take place gradually and therefore have limited effects on the health outcomes of populations in remote areas of Sub-Saharan Africa, UAV technologies have the potential to significantly accelerate access to health care, commodities, and data for providers and policy makers (Knoblauch et al. 2019). The introduction of UAVs can also catalyze discussions with governments on the costs and benefits of higher service levels, whether in faster and more frequent deliveries or in fulfilling emergency orders; for instance, it can prompt a conversation on how much governments and donors are willing to pay to avoid stock-outs of key commodities (ISG 2017). Zipline's commercial operations in Rwanda have demonstrated that health ministries may be willing to pay for UAV delivery in a fee-for-service model (USAID 2018).

At present, limited tools are available to help countries analyze and make informed decisions about how and when to integrate UAVs into public health programs and systems as part of a holistic approach to maximizing health and logistics objectives (Machagge 2018). Sustainable integration of drones into health systems requires in-country capacity and markets and businesses to locally own and operate a drone-supported system (Knoblauch et al. 2019). UASs have significant potential to improve the availability of health products in hard-to-reach locations. However, the economics and rationale for every potential use case must be considered individually, factoring in geography, UAS characteristics, and product and demand characteristics. The following factors are broadly indicative of a potential value-adding use case for UAS:

- High density of health facilities (within range of UAV)
- Health facilities difficult to access by road for large proportion of the year
- Products with high financial value, scarcity, or high health value (for example, lifesaving)
- Products with unpredictable demand at individual-facility level
- Products that have short shelf life, or are difficult to store at the last mile

Criteria need to be developed about whether a certain delivery (or set of deliveries) makes sense for the use of drone technology. To effectively use UAVs, it may also be important to consider layering use cases across programs to build flight numbers and thus lower the costs per km. The number of flights to each health facility can be increased by delivering or picking up a variety of cargo types. To answer the question of whether UASs add value for a particular use case, Wright et al. (2018) establish several basic rules for comparison. These rules lay out the methods that the UASs are compared with and the factors that are taken into consideration for the comparison (figure 1.11). However, a significant advantage of UAVs for at least one of the logistics objectives may be compelling enough to achieve overarching health objectives. In the case of safe blood for transfusion, speed, responsiveness, and risk reduction are paramount objectives, similar to those for diagnostic specimens. In contrast, availability and cost may be considered to be equally—or

FIGURE 1.11

Rules of thumb for identifying the advantages of UASs compared with traditional transport

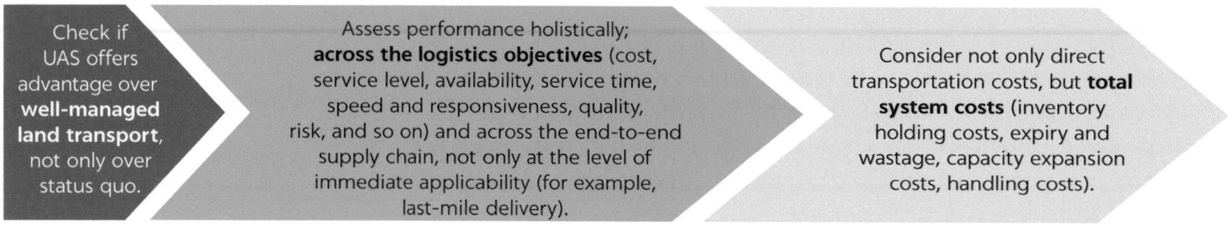

Check if UAS offers advantage over **well-managed land transport**, not only over status quo.

Assess performance holistically; **across the logistics objectives** (cost, service level, availability, service time, speed and responsiveness, quality, risk, and so on) and across the end-to-end supply chain, not only at the level of immediate applicability (for example, last-mile delivery).

Consider not only direct transportation costs, but **total system costs** (inventory holding costs, expiry and wastage, capacity expansion costs, handling costs).

Source: Adapted from Wright et al. (2018).
Note: UAS = unmanned aircraft system.

more—important objectives for vaccines and program and essential medicines (Machagge 2018).

Central storage combined with a cargo drone delivery strategy can overcome supply challenges in remote areas for a range of applications. This approach may be especially relevant for blood and antivenom, given the various blood types and the huge variety of antivenom, not all of which can be stocked at remote health care structures (Dirks 2017). Also the extensive research conducted by Wright et al. (2018) in East Africa suggests that the most effective approach to managing demand for long-tail products is to stock them at a central hub and provide them to individual health facilities when the need arises. This on-demand model aggregates demand across many individual health facilities, which makes it more predictable and allows lower buffer stock, and therefore saves money, especially on high-value commodities. This is known as the *inventory pooling effect*, commonly used in commercial systems with high demand variability across the system. On the other hand, there are advantages to establishing static high-frequency cargo drone routes for ensuring the regular supply of medical goods such as vaccines to small health care centers.

Emergency deliveries (or HIV or TB medicines) do not necessarily have to all be delivered by drone. Rather, a certain percentage of the medicines may present the most efficient use case. Another compelling case could be made for using UAVs to deliver vaccines when they are sent to pick up diagnostic samples (JSI 2018a).

According to a qualitative study with Médecins Sans Frontières experts on the relevance of drone-based transport, a payload of 10 kg and a range of about 100 km appears to be sufficient for the two most important identified drone applications: the small-scale and quick response to a specific limited need and the regular supply of remote health facilities (Dirks 2017). For specific applications in nonemergencies, such as the transport of medical samples, a payload of 1–2 kg would be sufficient. In addition to these promising applications, potential identified use cases according to the study include support of mass vaccination campaigns, transportation from central storage to remote health care centers, quick delivery of lifesaving basic medical supplies in a mass-casualty emergency response, and the regular support of routine telemedicine (the provision of health care from a distance). Furthermore, anesthetic drugs may be a good candidate for drone delivery in the event of natural disasters. For instance, small supplies for surgical applications could be delivered by drone to health facilities that are still functional after a disaster but that are running short on drugs to provide anesthesia.

Choosing between a fixed-wing and a rotary-wing aircraft requires that important trade-offs of range, payload, and the ability to hover and do vertical takeoff and landing (VTOL) be assessed. Although fixed-wing UAVs are more efficient and have longer ranges and greater payloads, rotary-wing drones take off and land vertically with the ability to hover during flight. For drone delivery of medicines in urban situations, rotor-based drones are more applicable, whereas rural locations benefit from the greater range associated with fixed-wing aircraft (Hii, Courtney, and Royall 2019). Most Médecins Sans Frontières experts interviewed by Dirks (2017) emphasized the need for a landing-type delivery rather than drop-off, given that sensitive items such as vaccines could break using a parachute drop. Drones that were capable of landing provided an advantage in the context of the immunization campaign implemented in the island nation of Vanuatu in 2018 because they enabled increased opportunities for community engagement, redistribution, and two-way deliveries (see box 1.1).

Drones are a technology solution that could help Tanzania's MSD reach its goal of 100 percent coverage, helping medicine reach challenging destinations, such as hilly areas (with steeper elevation slopes) or parts of the country where the primary and even the secondary road network is sparse. A similar argument

---

BOX 1.1

Integrating drones into Vanuatu's immunization supply chain

Since 2018, the use of drones in the vaccine supply chain has been in testing and validation phases in Vanuatu, an archipelago of approximately 80 islands in the Pacific Ocean. Only 20 of the 65 islands that are inhabited have airfields or roads, and many are only accessible by boat, for which access to fuel is difficult. Lack of a commercial shipping fleet makes scheduled sea transport to many islands very expensive, with transportation cost accounting for more than 90 percent of the average operational cost associated with vaccinating children ($20 per child) in 2018. Immunization services are provided at hospitals, health centers, and health dispensaries through fixed and outreach sessions; at aid posts (the most decentralized facility) in remote communities, they are provided by health workers who travel from cold chain–capable health facilities. In 2016, about 20 percent of Vanuatu's children were missing out on lifesaving vaccines such as against diphtheria, pertussis, and tetanus.

The target population for the 2018 immunization program was 8,998 children under age one across 36 health centers. With support from UNICEF, the Australian Department of Foreign Affairs and Trade, and the Global Fund to Fight AIDS, Tuberculosis and Malaria, Vanuatu's Ministry of Health and Ministry of Infrastructure and Public Utilities endorsed drone-based vaccine delivery trials in 2017 to reduce medical supply costs (from fewer losses, reduced expiry of medicines, and overall more efficient management of stock), increase flexibility of delivery and redistribution, reduce cold chain equipment costs by reducing the need for decentralized refrigeration, make more productive use of time for health workers (who would still need to travel to aid posts to administer vaccines but without the responsibility of transporting them from health centers), and take advantage of the possibility of transporting other lifesaving medical supplies along with vaccines.

In 2018–19, an open and public procurement process was implemented for three tenders to commercially deliver vaccines to communities on several islands. The Swoop Aero–operated drones landed at delivery sites, and nurses were trained to retrieve the payload. The landing of drones allowed for increased opportunities for community engagement, redistribution, and two-way deliveries. In contrast, the Wingcopter-operated drones did not land, instead delivering payload through a winch mechanism hovering at 10.5 meters above the delivery site, which limited deliveries to one-way only but enabled the rapid expansion of the number of delivery sites. Other commodities besides vaccines were also delivered (in the one-way scenario) and redistributed (in the two-way scenario), including syringes, antibiotics, oxytocin, dental supplies, and others. Over nine weeks, 1,066 women and children were immunized with vaccines delivered by drone, and it was concluded that drones can safely deliver temperature-controlled medical supplies to last-mile communities, with the landing (two-way) system found to be preferable because it addresses both supply streams given that health workers are able to send a vaccine arrival report directly back with the drone after reviewing freeze-tags and vaccine vial monitors. The conclusion was that a large number of flights are required for drones to reach a competitive tipping point, and that it would be preferable to widen drone services beyond the health sector to optimize equipment usage.

Source: UPDWG 2019.

can be made for drone integration in a number of other countries in the region, as demonstrated by maps 1.7 and 1.8.

Certain regions in Tanzania, including Kilimanjaro, Mwanza, and Shinyanga, are expected to be significantly more affected by extreme fluvial flooding in future decades as a result of climate change, as highlighted in map 1.9 (Oxford Infrastructure Analytics Ltd. 2018), which may make road-based medical goods deliveries even more challenging than at present. Supplementing existing road, rail, and waterway delivery, drones can accelerate delivery time and reduce costs. When time is of the essence, as is often the case when delivering antivenom or blood for transfusions, the adoption of UAVs could potentially save thousands of lives. For a few years, WeRobotics Flying Labs has been building the local capacity in Tanzania that will ultimately be required to operate and maintain UAVs in-country (USAID 2018).

Based on previous research, the Mwanza-Nansio corridor is considered to be a *potentially good candidate for high-frequency drone delivery, with secondary* service from Nansio to outer facilities. The Deliver Future / Drone X project, funded by the German Agency for International Cooperation and piloted between January 2017 and May 2018, was conceived of as an appropriate response to Mwanza's medical supply chain inefficiencies and was implemented in close cooperation with the Regional Commissioner of Mwanza; the Ministry of Health, Community Development, Gender, Elderly and Children; MSD; the Tanzania Civil Aviation Authority; and the Dar Teknohama Business Incubator. As part of the project, a long-distance (64-km) air corridor (operated exclusively

MAP 1.7

Slope

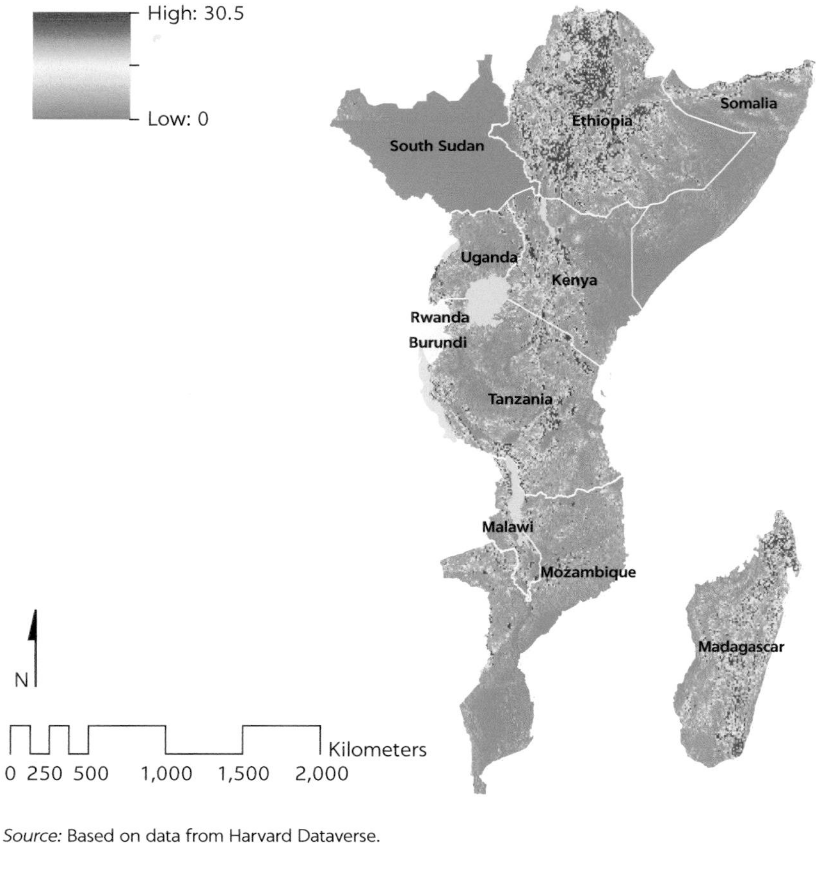

Source: Based on data from Harvard Dataverse.

MAP 1.8

Primary and secondary road coverage

—— Primary and secondary roads

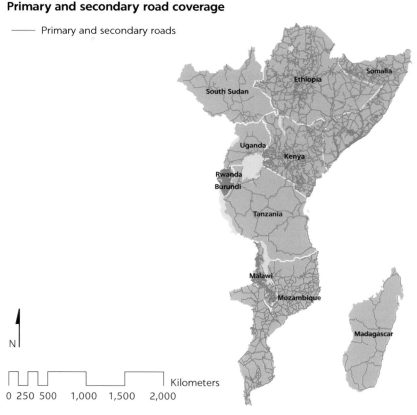

Source: Based on Global Roads Inventory Project (2018) data.

by Wingcopter[11]) was established, along with stakeholder networks across Mwanza (Rabien 2018). Analysis by JSI (2018a) produced concrete recommendations for the priority health products relevant for the Drone X business case in Mwanza. The analysis assumed that deliveries would be made from the supplier (MSD in most cases) to facilities and not from a potential drone port in Ukerewe. JSI identified deliveries of HIV VL tests and other laboratory diagnostic samples as the top priority, followed by vaccine delivery from the district level to the health facilities.

The use of drones in the medical goods delivery market in other countries in Sub-Saharan Africa is at various stages. For example, for immunization supply chains, drone use in 2019 was in the "advocacy and testing" phase in the Democratic Republic of Congo but had reached the "expanding impact" stage in Ghana and Rwanda (UPDWG 2019).

Rwanda boasts the world's first one-way airdrop medical cargo drone delivery service operated by the international health logistics company Zipline, which since 2016 has been delivering sachets of blood to the country's hospitals thanks to an agreement with the government that gave its drones the status of government flights. Zipline works directly with the Ministry of Health and the Ministry of ICT and Innovation through long-term (two or more year) service contracts; its system is integrated into Rwanda's supply chain system. The distribution center and drone port are in Muhanga District of the Southern Province, colocated with a medical warehouse to maximize speed and product availability. The Zipline cargo drones are operated out of operations centers the size of a shipping container that are placed close to the public medical

MAP 1.9

Predicted regional changes in fluvial flooding in Tanzania due to climate change

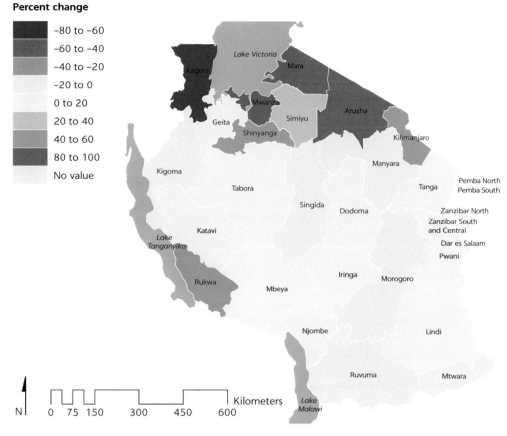

Percent change

- −80 to −60
- −60 to −40
- −40 to −20
- −20 to 0
- 0 to 20
- 20 to 40
- 40 to 60
- 80 to 100
- No value

Source: Oxford Infrastructure Analytics Ltd. 2018.
Note: Map shows the percentage change between the predicted mean regional flood areas in the future and current flood model outputs across all models and all return periods.

warehouses. Each operations center has a service radius of up to 75 km, with the drone carrying up to 1.5 kg of cargo that the UAV drops into a marked "mailbox" area using a parachute before it returns without landing to the operations center. The UAVs can land on the equivalent area of two parking spaces and are launched with a catapult. Each drone is estimated to make about 15 deliveries per day (USAID 2017). As of 2018, the project had expanded from the initial 21 district hospitals to all of Rwanda, including dozens of hospitals and approximately 500 small rural clinics.

Since its entry into the Rwandan market, Zipline has also been in talks with other countries in Sub-Saharan Africa to offer drone services for the delivery of blood and other medical goods. Since 2019, Zipline has been operating in Ghana, delivering medicines and vaccines. In the spring of 2020, in the context of the COVID-19 pandemic, the company further expanded its dialogue with the Ghanaian government to identify opportunities for drone applications. The applications that have been launched as a result of the talks include covering a share of the needed vaccine deliveries by UAV to support overall "social distancing" principles (that is, reduce in-person deliveries involving human drivers or pilots), and ensuring on-demand deliveries of personal protective equipment to frontline health care workers. In addition, plans are for UAVs to be used in the delivery of COVID-19 swabs back to testing labs using the drone system already in place. This project will enable the transport time to be reduced from the current 4–5 hours (for trips from rural health centers to testing labs) to less than an hour (from rural health centers to the Zipline operations centers, where the samples will be received, packaged, and from there flown to the testing labs). The pandemic-related UAV activities have been approved by Ghana's Civil Aviation Authority and are consistent with WHO guidelines for protecting the integrity of the COVID-19 samples. The case of Zipline in Ghana demonstrates the benefits of having a countrywide drone network in place in the event of a pandemic, not the original goal but a by-product of having a diversified, robust supply chain.

Although Zipline's one-way medical goods drop-off systems remain the few fully operational such projects in Sub-Saharan Africa, several pilot projects are being conducted in other countries exploring *bidirectional* transport—the ability to land at a remote health facility or a village and return (Knoblauch et al. 2019) in addition to one-way deliveries. In Malawi, UAV pilot initiatives have been conducted by the Ministry of Health and Population, Malawi Blood Transfusion Service (MBTS), and the Pharmacy Medicines and Poisons Board, with external partners such as VillageReach and the United States Agency for International Development (USAID) Global Health Supply Chain Program–Procurement and Supply Management (GHSC-PSM) program. In conjunction with the Ministry of Health and Population and UNICEF, in 2017 Malawi's Department of Civil Aviation (DCA) opened a drone-testing corridor in Kasungu in central Malawi, which remains the only dedicated unmanned flight-testing space for humanitarian purposes in East Africa.[12] With a radius of 40 km, an allowed operational ceiling up to 400-meter altitude, a total area of 5,024 km², and a population catchment of 650,000 people, the corridor is designed to provide a controlled high-altitude, long-endurance flight-testing platform for the private sector, institutions, nongovernmental organizations, and other partners to explore how drones can be used in humanitarian and development applications. The corridor is centered around a local airfield with a 1,200-meter runway and serves one central hospital laboratory and more than 300 schools, health centers, and clinics (Juskauskas 2019). After a period of testing at the corridor, companies are evaluated by DCA for flights flown elsewhere; on this basis, DCA reserves the right to approve or deny UAV operation in public air space (Department of Civil Aviation et al. 2019).

A medical supplies delivery pilot in Kasungu was implemented based on initial scoping missions that identified the villages that tend to be particularly isolated, either because of severe flooding during the rainy season or because of their location in Lake Malawi. Only about 78 percent of women and 66 percent of men in the Kasungu District have ever been tested for HIV and received results (compared with 87 percent and 76 percent, respectively, in Lilongwe City), making it a good testing ground for drone-based deliveries of lab samples. However, Kasungu District was selected for the drone pilot not only because of its medical needs and natural characteristics but also the favorable regulatory environment. The initial project focuses on the collection of medical samples (for both HIV and TB diagnosis) and delivery of medication, with deliveries by landing taking place bidirectionally at district hospitals, peripheral health centers, and blood testing sites.

Since 2018, the government of Malawi and UNICEF have also worked together to use drones to reduce the spread of malaria by identifying breeding grounds for mosquitoes transmitting the disease and to combat cholera. Following the progress of the drone outreach program in Malawi, UNICEF

Namibia, too, is looking to deploy drones to transport blood samples from rural areas in the Zambezi to central laboratories (Adeshokan 2019).

In the USAID GHSC-PSM project in Malawi, the drone launches from Nkhata Bay District Hospital (map 1.10), allowing pharmacy and laboratory staff to participate in operations on a day-to-day basis. At the smaller and more remote Chizimulu Health Center, the project team trained local staff in drone flying, proper loading and unloading, communications procedures, and securing the landing area, so local staff are fully capable of receiving and dispatching the drone without assistance (Aryal and Dubin 2019). The project was conceived based on an initial assessment by JSI (2018c), but a full demand assessment for various medical goods has not yet been conducted.

Testing of drones in Malawi's health programs has also been supported by VillageReach. The organization's work on drone-based medical deliveries in East Africa began in 2015, and a costing study for delivery of laboratory samples

MAP 1.10

Health facilities in Malawi

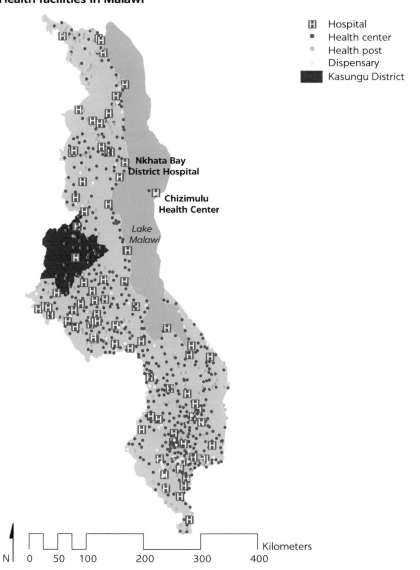

Source: Data provided by UNICEF.

in Malawi was completed in 2016. The first pilot was launched in 2016 to gain an understanding of whether drones could speed up EID for HIV. In 2017–19, the scope of the initiative was expanded to identify opportunities for using drones in the delivery of injectable oxytocin and blood for transfusion during maternal health emergencies, given the urgency of the maternal hemorrhaging situation in the country, the weak requisite infrastructure to rapidly respond to life-threatening incidents, and the weak storage capacity in rural district hospitals and health facilities (Matemba 2019). In this use case, VillageReach worked with the Ministry of Health and Population, MBTS, the Pharmacy Medicines and Poisons Board, and NextWing Corporation.

The methodology included key informant interviews and simulated transportation of blood and oxytocin from MBTS in Lilongwe to Lilongwe and Dowa Districts, including seven blood transfusion facilities. Three hybrid (fixed-wing and quadcopter) autonomous but monitored drones are planned to be used, with maximum flight range of 80 km and maximum payload of 1 kg. As in the HIV and TB testing pilot project implemented by UNICEF, the VillageReach initiative involves bidirectional transport delivery with landing at the various types of health facilities.

Drones on a pilot scale have been introduced in the region to test their applicability in telemedicine—medicines are delivered directly to the patient instead of the patient being asked to take a long and potentially dangerous journey to receive the medication. In the remote Androrangavola commune in Madagascar, a Drone Observed Therapy System (DrOTS) for TB diagnosis and treatment, shown schematically in figure 1.12, was piloted in July 2016 through a partnership between

FIGURE 1.12

Drone Observed Therapy System (DrOTS), Madagascar

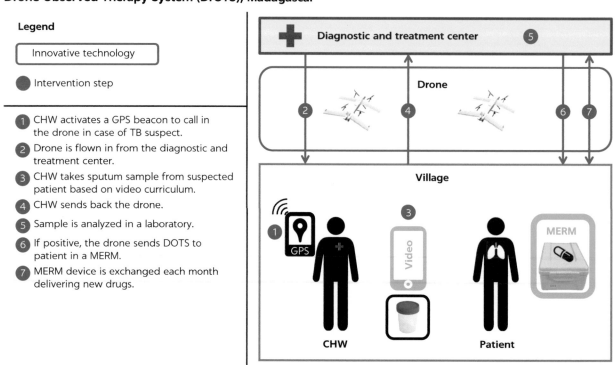

Source: Astrid Knoblauch, Drone Observed Therapy System in Remote Madagascar (DrOTS) Project, in Small (2017).
Note: CHW = community health worker; DOTS = directly observed therapy short course; GPS = global positioning system; MERM = medical event reminder monitoring; TB = tuberculosis.

Stony Brook University, Institute Pasteur Madagascar, and the National Tuberculosis Control Program of the Ministry of Health. The goal of the program, which targeted 35,000 people, was to combine the drone capabilities with technologies improving patient education and symptom monitoring (Small 2017). GPS-guided hybrid (fixed-wing and quadcopter), fully autonomous drones with a maximum range of 60 km and a payload of 2.2 kg, capable of carrying 160 lab samples or four lifesaving blood units, were used to transport TB lab samples from rural villages to Stony Brook's Center ValBio research station on the edge of Madagascar's Ranomafana National Park, where the samples could be properly stored and analyzed for TB diagnosis (USAID 2017). The average distances flown by the drone were 15–20 km each way, and all technology aspects of the drone pilot were managed from a central facility, which included a launching pad and a warehouse for storing spare parts and employed one lead technician (a US-trained engineer with flying permits) and three Malagasy trainee technicians. Once a day, the drones returned to the facility for overnight maintenance. Although drones were not fully integrated into the local health system, proof-of-concept was achieved. The Malagasy government, major health service providers, and respective funders now plan to integrate drones into their health provision activities in Madagascar (Knoblauch et al. 2019), with the original drone pilot program now redirected to malaria rather than TB because of a perceived larger potential impact.

In Mozambique, VillageReach has explored using drones for transporting TB samples to laboratories and the timely delivery of test results to initiate lifesaving treatment and prevent disease transmission. Mozambique's TB incidence is by far the highest in the East Africa region, at 551 cases per 100,000 population in 2018, followed distantly by Kenya and Somalia with 260–290 cases. The effective coverage of TB treatment in Mozambique is only 40 percent (figure 1.13). In collaboration with Mozambique's Ministry of Health and the Instituto Nacional de Saúde, Mozambique's National Institute of Health, VillageReach launched the Unmanned Aerial Systems for Tuberculosis project in Mozambique in 2018. UAV flights from the central National Institute of Health lab to a health facility in Maputo Province and back intended to demonstrate the capability of UAVs to maintain the quality of TB samples for diagnosis and generate evidence on the associated costs and performance related to the sample referral network (VillageReach 2019a).

FIGURE 1.13

Tuberculosis incidence and treatment coverage

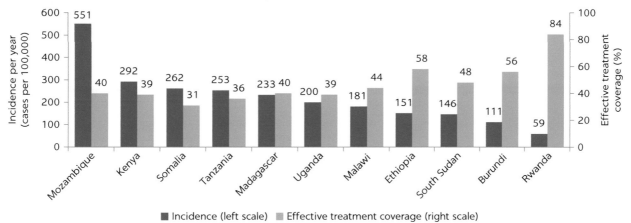

Source: Data from WHO (2018).

In the Democratic Republic of Congo's Equateur Province, a partnership between the Ministry of Public Health (national and provincial), the Civil Aviation Authority, Gavi, and VillageReach, called Next-Generation Supply Chains, is being implemented to improve the supply chain and product availability, specifically demonstrating two-way transport of vaccines and other medical commodities using drones. VillageReach has been working in the country for several years with the government of the Democratic Republic of Congo and the WHO Expanded Programme on Immunization to explore solutions for improving the immunization supply chain. A major change in the system was introduced in 2015–19 following a 2014 evaluation (UNICEF and WHO 2014), introducing more direct deliveries of vaccines from the provincial level to health centers. Although this resulted in large impacts for reducing stock-outs, some areas of the province remained difficult to serve because of transportation challenges, providing the rationale for piloting the integration of drone-based solutions. The introduction of the drone program has involved significant up-front preparatory work by VillageReach, the Ministry of Public Health, and other stakeholders to create the enabling environment, including the introduction of a dedicated review commission and an extensive review of the drone platforms that could meet government requirements for a variety of medical use cases. Following a competitive global selection process, in the summer of 2019, the drone company Swoop Aero conducted a series of demonstration drone flights to deliver vaccines, syringes, medicines, and other supplies from the city of Mbandaka to the village of Widjifake (an 80-km round trip), flying across the Congo River, forests, and the populated area of Mbandaka (VillageReach 2019a). Phase 2 of the program will entail at least 12 months of flights covering all of Equateur Province, starting with vaccine transport but eventually expanding to other medical goods, with the eventual goal of structuring a sustainable public-private partnership–based business model.

Potential time savings and impacts on quality

In addition to cost and quality, increasing the speed of response is a critical objective for deliveries of goods such as lab samples and blood for transfusion. Sputum samples must be tested within 72 hours of the sample being obtained; the transit time for blood and blood components should not exceed 24 hours (Global Laboratory Initiative 2014). For DBS specimen transport, the primary goal is to get HIV patients on appropriate treatment plans as quickly as possible (Phillips et al. 2016). The delivery of blood is often needed in life-threatening situations, so saving even a few minutes could be of great value (Wright et al. 2018).

In these use cases, small, fixed-wing UAVs can offer speed and responsiveness advantages over land transport because UAVs are not affected by road circuity, and small fixed-wing UAVs (that travel about 100 km/h) are much faster than motorcycles (about 40–50 km/h). For less-predictable medical goods demand categories (Rh-negative blood, plasma, and so on), analyses of small fixed-wing UAVs for blood delivery in Rwanda and Tanzania have shown that, compared with transport by land cruiser or van, UAVs could reduce delivery time by a factor of four (Wright et al. 2018).

Cold chain items such as lab or blood samples, and especially vaccines, are the most challenging medical goods to transport because of the need to continuously maintain the cold chain. Whole blood must be maintained at a temperature between 2°C and 10°C between the collection site and the laboratory

(WHO 2005). Sputum samples must be kept at a temperature between 2°C and 8°C (Global Laboratory Initiative 2014). Other medical goods, such as medicines, must generally be kept at temperatures less than 25°C, which may be challenging in many parts of Sub-Saharan Africa.

Amukele et al. (2017) tested whether drone transportation of blood in the United States would have any impact on its quality—regular chemistry, hematology, or coagulation. Half of the samples were held stationary, while the other samples were flown for three hours (258 km) in a custom active cooling box mounted on the drone.[13] The study found that samples on drones yielded very similar results to those transported terrestrially. The authors conclude that long drone flights of biological samples are feasible but require stringent environmental controls to ensure consistent results. In a similar study, Hii, Courtney, and Royall (2019) tested the impact of drone transportation, specifically the effects of temperature and vibration, on the quality of insulin, one of the most important lifesaving drugs for treating type I diabetes. They flew the medicinal product, Actrapid, in its original packaging, using a small, commercially available drone. The results showed that drone transportation had no adverse impacts on Actrapid, providing evidence that drone transportation of insulin is feasible from the point of view of pharmaceutical stability and maintenance of medicine quality.

Encouraging initial results have also been recorded as part of the Wingcopter-operated drone delivery trials (Phase 1) in Mwanza, Tanzania. Performing approximately six flights per day, the drone-based trial provides evidence of the feasibility of reducing delivery times for essential and emergency medicines from 1.5–3.0 days to 1 hour. The delivery times for blood samples from health facilities to the Bugando referral hospital were cut from 3–5 days to 1.2 hours, with blood sample results transmitted back to health facilities in less than six hours (Mulamula 2018).

In Rwanda, drones deliver blood to regional hospitals from a base in the eastern part of the country. Health workers at clinics and hospitals send their orders online or via text message, and within minutes the orders are launched using fixed-wing drones, powered by two electric engines and moving at a speed of more than 100 km/h (Zipline 2018). For the clinics the average delivery time for routine medical supplies in Rwanda has thus been reduced from four hours to 45 minutes (McVeigh 2018). The company reported in 2019 that its service is, on average, 82 percent faster than conventional transport methods.

In 2018, an assessment of the sample referral transportation network and health supply chain in two districts of Malawi with hard-to-reach facilities was commissioned by UNICEF. The assessment modeled the benefits of integrating drones into an optimized specimen referral system. Because the specimen referral system requires reverse logistics ability, that is, transport of items from the health facilities back to the district, this analysis examined UAVs with VTOL capability for Nkhata Bay and Likoma Island districts. The long distance and aerial conditions of crossing the lake to Likoma Island present a different set of equipment requirements than those for Nkhata Bay, so the analysis assumed that UAV equipment with the appropriate features and capabilities would be used for these segments (JSI 2018b). The study (JSI 2018c) finds that the benefits included increased equity and access for patients, responsiveness to urgent needs, and potential use in emergencies and catastrophes, such as floods. Complementing the insights provided by these mostly modeling-based studies, the pilot project operated by USAID's GHSC-PSM has generated empirical

---

BOX 1.2

Malawi drone corridor: An opportunity to assess longer-term impacts

The United States Agency for International Development's Global Health Supply Chain Program–Procurement and Supply Management (GHSC-PSM) in Malawi is piloting drones as a last-mile delivery solution for the country's health supply chain. Since July 2019, the project's drone has been running daily cargo flights between Nkhata Bay District Hospital on the lakeshore and St. Peter's Hospital on Likoma Island. The drone carries critical health commodities and patient samples back and forth, shortening transportation times by weeks and ensuring samples are processed and results shared with clinicians in a timely manner. Since routine flights began, the GHSC-PSM drone has flown more than 2,600 miles, reliably delivering diagnosis and treatment eight weeks earlier than the previous time. Over a period of just three months in 2019, GHSC-PSM completed 56 cargo flights, transporting more than 100 viral load, early infant diagnosis, and tuberculosis samples for testing, as well as micropipettes for HIV testing, emergency medicines, vaccines, and syphilis test kits. Unlike short-term pilots, sustained bidirectional flights require intimate knowledge of the in-country health system and supply chain, including the subsystems for lab referrals and reporting. Because the drone flies over Lake Malawi each day carrying lab results and medical cargo, the project gathers valuable information that leads to a better understanding of the impact of drones on health outcomes. The project completed sustained bidirectional flights for several months, providing a relatively long-term learning opportunity.

Sources: Adeshokan 2019; Aryal and Dubin 2019.

---

evidence on the actual advantages and challenges associated with drone-based transport of medical goods in Malawi (see box 1.2).

The ongoing work by VillageReach in Mozambique, supported by the UK Department for International Development, on the use of drones for TB sample transport will provide insights into how UASs can affect both the availability of and turnaround time for TB lab services at the last mile, thus shaping how UASs are used for strengthening laboratory networks.

In the VillageReach immunization project in the Democratic Republic of Congo's Equateur Province, the use of drones in lieu of road transport cut delivery time from roughly three hours to 20 minutes one way, with 470 children immunized over a span of five days across five rural areas (UPDWG 2019). The test flights also confirmed that cold chain could be maintained using UAVs.

However, it is worth considering the entire diagnostic and treatment cascade to understand whether transport is the major bottleneck for turnaround time of diagnostic results. There are many potential sources of delay within the supply chain, including at the health facilities (because of batching, or when specimens sit at various locations within a facility with no coordination with when transport is available), at the laboratory (because of equipment breakdown or reagent stock-out, or batching), and delay in results return. For example, UNICEF Malawi funded a study that found that geographical constraints, infrequent transportation of samples, and insufficient staff and testing facilities all delayed turnaround time of medicine in the two studied districts (UNICEF Malawi 2019). Therefore, a decrease in transportation time potentially offered by a drone would only affect one part of the cascade, which may offer only a small reduction in overall turnaround time (Wright et al. 2018).

Potential cost savings

Concurrent advances in engineering and manufacturing technologies are pushing the pace of change and future drone capabilities and costs (ITF 2018). However, despite fast technological development in the drone-technology sector (propulsion, payload, endurance, stability), data and experience are lacking in some areas, such as on the total costs of drone ownership (Müller, Rudolph, and Janke 2019).

In addition to the cost of the UAV itself, up-front capital investment costs are associated with the needed ground equipment:

- Buildout of storage or warehouse and launch and landing area, if required
- Equipment such as laptops, tablets, screens, transmitters, and so forth
- Office and work area, including desks

For small hybrid (battery-operated), multicopter, and small hybrid (gas/electric) UAVs, JSI (2018c), in a study focused on the Nkhata Bay in the Northern Region of Malawi, estimated these additional capital investment costs at $30,000. For the medium hybrid UAV the estimate was significantly higher, at $50,000. The estimated capital investment needs for the UAVs themselves ranged even more widely, from $6,000 for a multicopter and $30,000 for a small hybrid (battery-operated) to $85,000 for a small hybrid (gas/electric) UAV and $100,000 for a medium hybrid.

In addition to capital investment costs, UAV annual operating costs include components such as ground and operator staff, insurance, fuel and electricity, utilities, batteries, spares, maintenance, and connectivity.

Hassanalian and Abdelkefi (2017) present an extensive overview of drones by weight, wingspan, payload capacity, flight range, and flight endurance as well as by selected propulsion system and mission capabilities, such as VTOL. Different drone types' characteristics are also summarized in Wright et al. (2018), suggesting that total costs per ton-km are lowest for large fixed-wing drones, followed by hybrid-medium and small fixed-wing drones. The heterogeneity of drone designs is driven by the large variety of potential drone use cases, with specific use cases leading to the specification of drone mission parameters such as intended payload (weight and dimensions), range, altitude, and speed (ITF 2018). According to Wright et al.'s (2018) research across several countries in East and Southern Africa, an average facility can comfortably fit a week's worth of specimens in a single UAV flight (even for the smallest VTOL UAVs). However, the cost advantages are not clear for using multicopter and hybrid UAVs because they are more expensive on a cost-per-km basis than fixed wing, more expensive than motorcycles, and have a shorter range than fixed-wing UAVs. If motorcycles take a multistop route, they are in fact even cheaper on a per-facility basis, often by a factor of four or more depending on facility density, road network, and estimates of route costs. Therefore, it will be extremely difficult for UASs to be cost competitive with motorcycles on multistop routes.

The cost-effectiveness of drones compared with traditional transport systems, however, can vary significantly with the assessment methodology used. Although a strict comparison of UAV cost to motorcycle cost will, in most cases, tilt in favor of the motorcycle, including the cost of fuel and the human resources required to operate a vehicle or motorcycle for hours compared with a shorter UAV flight (with a single person operating multiple drones) may deliver different findings. Different UAV systems will also have different human resource cost implications: a fixed-wing drone that does not land has limited training costs because trained staff are only needed at the "nest" (the original take-off location to which the drone returns after dropping off the goods), whereas a UAV that lands has much higher training costs because all facility staff need to master operation of the drone. These types of indirect costs are extremely difficult to assess (USAID 2017). In the absence of an actual cost track record, modeling can provide valuable input for decision-makers (Phillips et al. 2016).

The Médecins Sans Frontières logistics managers interviewed by Dirks (2017) acknowledged the value of using cargo drones in countries where access is complicated; they believe it will be cost-effective in places where UAVs could replace road transport that takes several days. In addition to the Central African Republic and Mali, parts of which are inaccessible because of security issues and extended flooding, Kenya's Homa Bay Ndhiwa region in the southwestern part of the country was highlighted as a setting in which drones could generate cost savings for the delivery of blood samples for HIV and TB testing. Drone-based transportation was hypothesized to possibly be cost competitive in Kenya, Malawi, and Uganda for transportation of samples, medicines, and vaccines, and in Yida settlement and the Nuba Mountains, South Sudan, for vaccinations and HIV and TB treatment for refugees from Sudan.

A Microsoft Excel–based simulation tool[14] developed by JSI in 2018 and updated using new data as part of the current study (table 1.5) allows first-pass screening for individual programs and countries to be conducted to rapidly prioritize potential use cases for UAV delivery, based on the programs' and countries' unique situations, and to approximate relative costs compared with land transport and potential benefits. The analysis suggests that most UAVs are still not transport cost–competitive with motorcycles. However, the case for UAVs can be made on *total* system costs and benefits (Machagge 2018). A UAV Delivery Decision Tool[15] has also been developed by FHI 360 to help clearly define promising use cases and to help select testing sites where UASs may address transport challenges. The development of possible business models for drone applications in medical goods delivery is underway in several countries, with box 1.3 illustrating the case of Ghana.

TABLE 1.5 **UAV characteristics and associated costs**

CHARACTERISTIC	MULTICOPTER	SMALL FIXED WING	SMALL FIXED WING; VERY LOW FIXED COST	LARGE FIXED WING (NEEDS RUNWAY)	HYBRID, SMALL	HYBRID, SMALL (ALTERNATIVE FUEL)	HYBRID, MEDIUM
Range (km)	20	150	70	500	85	100	100
Weight capacity (kg)	2	1.5	1	100	2	4	20
Volume capacity (liters)	10	10	0.4	230	8	13	25
Reverse logistics capability (VTOL)	Yes	No	No	No	Yes	Yes	Yes
Estimated number of flights per year	1,511	111,562	362,122	11,806	27,294	23,611	11,806
Estimated total (fixed and variable) costs per flight given flights/year ($)	35.17	14.47	13.05	185.70	36.74	90.00	125.95
Estimated total costs per km given flights/year ($)	2.64	0.14	0.28	0.55	0.65	1.73	1.88
Total cost/ton-km ($)	1,319	96	278	6	323	433	94
Total cost/m³-km ($)	264	14	696	2	81	133	75

Source: JSI Tool for Determining Cost Effective Use Cases for Autonomous Aerial Systems, flight estimates based on Ukerewe District, Tanzania, context.
Note: UAV = unmanned aerial vehicle; VTOL = vertical takeoff and landing.

BOX 1.3

Dr.One drone-based medical goods delivery business model (Ghana)

The Dr.One Proof of Concept is a joint project of the Ghana Health Services (GHS), the United Nations Population Fund, the Netherlands Aerospace Centre, and IDI Snowmobile B.V. Dr.One aims to improve health care in remote areas of developing countries by using small drones for the transportation of medical goods. The Dr.One concept is designed to operate as part of community-based health care systems, in compliance with all rules and regulations, including those for civil air space.

The Dr.One operator organization uses a *usage-based* model; a small fee is charged to the health supply chain for every delivery made. In addition, the operator organization generates revenue through the *advertisement* model, by providing targeted digital content at remote locations. An operator organization can operate as part of a national Dr.One organization or can be integrated within existing health care organizations, but can also be *franchised* out to contractors. The production and maintenance organizations initially start off with models that focus on paying for products and hours spent, and as the concept matures, migrate toward a *usage-based* model. The training organization uses a *one-time up-front charge plus maintenance* model and a *subscription* model for continued access to training information.

Source: Drones for Development 2016.

A highly beneficial aspect of airborne transportation is its use of line-of-sight, "as the crow flies," connection. Furthermore, no spending on supporting infrastructure en route is needed. This attribute may be very important for the last mile of delivery, which is often the most expensive part of the transport cost of medical goods such as vaccines, accounting for 30–40 percent of total supply chain costs (USAID 2013). Roads and rail tracks must be maintained for overland transport, and every new connection between two points needs significant investment. For airborne delivery, there is only the need for loading, departure, reception, and unloading (Müller, Rudolph, and Janke 2019). However, the local context is a significant factor in any assessment of the potential cost savings drone-based transportation can deliver. For example, in Rwanda, the relatively mountainous terrain makes drones an attractive option, but in countries with better land transport infrastructure and a relatively flat landscape the cost and time advantages of drones versus traditional transport modes may be more muted.

If overall infrastructure investment costs, not just operating costs, are considered, drone-based deliveries can provide a cost advantage in areas where substantial investments would be required to achieve basic rural road accessibility. Mikou et al. (2019) estimate that transporting weekly medical and school supplies to remote areas in Sierra Leone by drone would cost three times less every year than increasing the Rural Access Index from 30 percent to 31 percent, and 22 times less than increasing the index from 69 percent to 70 percent.[16] Moreover, these costs are likely to decrease quickly as drone markets mature and the number of service providers increases. The authors argue that drones can therefore be a useful short-term solution to improve the welfare of remote populations.

Similarly, following testing in Lesotho, the California-based company Matternet estimated that it would cost $900,000 to operate 50 base stations and 150 drones ($0.24 per flight) throughout the capital city of Maseru. The authors suggest that such a system would be more cost-effective than building roads, estimated at $1 million for a 2-km one-lane road (Maisonet-Guzman 2014).

Another significant advantage of drones in medical goods delivery is the reduced reliance on peripheral forecasting, given that just-in-time orders can be placed by text. A related potential benefit is the reduced need to store expensive products or products requiring cold chain at the peripheral (remote health post) level, with associated reductions in the likelihood of expiry and lower overall storage costs (ISG 2017). For example, the rising cost of vaccines makes it necessary to maintain lower stock levels, reduce wastage, accurately forecast needs, and avoid equipment breakdowns (UNICEF and WHO 2014). In other words, the use of drones and the resulting time savings can also deliver significant non-transport cost savings. Vaccines and other medical goods that are compromised because of excessively lengthy transportation result in many thousands of wasted dollars and also lead to distrust by the receiving population (WHO 2015).

For certain use cases and route typologies, locally made, low-cost drone technologies may offer a feasible solution and cost savings compared with land-based modes. For example, Micromek is a Malawi-based hardware start-up producing low-cost UAVs that will serve the health care and environmental-monitoring communities in Malawi and across Africa. By working hand in hand with the Virginia Tech University Unmanned Systems Lab, the company developed a low-cost UAV (EcoSoar) to provide a lower-cost delivery option and improve delivery reliability of medical goods. EcoSoar is used to deliver remote medicine, DBS specimens, vaccines, and other critical medicines from clinics in the UNICEF Kasungu drone corridor in Malawi to the Kasungu District Hospital.[17]

Limited existing track record on actual operating costs

Recent reports on UAS adoption document use cases in a developing country public health context, summarizing technical use cases and identifying challenges; however, there is still little actual flight experience and data on potential cost savings. In most East African countries where drones for deliveries are operated at all, only small-scale operations have taken place so far, and the track record on operating costs remains limited.

In 2019, Zipline reported having completed 8,000 deliveries in Rwanda, transporting more than 15,000 units of blood, suggesting that its service had helped save millions of dollars through expiry reduction, improved primary care, reduced referrals, elimination of emergency road deliveries, and so on. Zipline reports that it can deliver medicines to health facilities four times as frequently compared to what is feasible with overland transport modes for the same price—or less. The drones' delivery system costs less than it would cost to make the same delivery using a motorbike, which was previously the only convenient means for an emergency supply run, given that drones can deliver medical supplies without incurring costs such as driver and health worker time and wages, vehicles, fuel, storage of packet red blood cells on site at health facilities (utilities, equipment, value of inventory), and blood units that go to waste when they expire (including $100/unit to test and type the blood). According to Zipline's service contracts with the Ministry of Health, price per delivery varies based on the volume of deliveries and delivery context but is generally on par with current delivery costs (USAID 2017). The company reports that a $2 million investment is typically made in a drone facility, at the same time creating about 100 local jobs. However, little detailed data are available on the company's operating costs;[18] Zipline designs, manufactures, and operates the entire system in-house.

Affordability and cost-effectiveness were also evaluated for the Madagascar DrOTS pilot that was launched in 2016 (Small 2017). Evaluation criteria included the total cost of DrOTS in relation to the cost of other treatment regimens and the total cost of treatment, as well as comparative cost-effectiveness—incremental cost per death and disability-adjusted life year (DALY) averted (only TB focused)—of the DrOTS standard of care in remote settings. The reported cost breakdown of the initial pilot was as follows:

- One-time, up-front: $33,000 for technology procurement (three drones, batteries, customs, and taxes) to be used over five years and $9,300 in training costs for all local staff on the ground
- Annual: $3,000 for drone insurance; $3,000 for drone maintenance, software, GPS, and so forth; and $3,000 for the maintenance of other infrastructure

The preliminary cost-effectiveness study for the initiative found that, given TB prevalence rates in Madagascar, the incremental cost-effectiveness per DALY averted was $1,040, representing the additional investment in drone transport compared with traditional transport required to gain one DALY. Although the pilot was therefore considered cost-effective according to WHO guidelines, it was not deemed to be an affordable solution in Madagascar's context.

Current work by VillageReach in Malawi includes documentation of costs of a UAV-supported transport system for blood and uterotonic drugs for emergency cases as compared with the current blood and uterotonic drugs transport system. Similarly, in Mozambique, the ongoing VillageReach initiative will allow the costs of using UAVs in TB sample transport to be documented. Collection of these data will enable the best UAV use cases for TB sample transport to be identified and potential business models for sustaining the use of UAVs for laboratory sample transport to be developed.[19]

In addition to the limited number of ex post evaluations of existing drone pilot programs, several *modeled* cost impact studies have been conducted in parallel with or in anticipation of actual drone trials.

Review of existing studies in Sub-Saharan Africa

Numerous modeled cost-competitiveness studies have been implemented by other development organizations in Malawi. In conjunction with the UAV field tests organized by UNICEF, the government of Malawi, and Matternet in March 2016, VillageReach conducted a costing study in Malawi (Phillips et al. 2016) using a Microsoft Excel–based cost modeling tool[20] it developed, in which drone costs were compared with the standard method of transporting lab samples and results between health facilities and laboratories via motorcycle. Analysis focused on the Lilongwe District, and more specifically, the subset of service delivery points that send DBS samples to Kamuzu Central Hospital via motorcycle transport. The study analyzed costs under four scenarios:

Scenario 1 (loops). Transport of EID DBS samples using loops, in which a vehicle visits multiple service delivery points before returning to the laboratory.

Scenario 2 (loops). The same transportation loops as in scenario 1, but for both VL and EID DBS samples.

Scenario 3 (hub and spoke). EID transport for only the service delivery points within 25-km UAV flight range of the laboratory. Vehicles (motorcycles or

UAVs) used hub-and-spoke routes to travel from the laboratory to each service delivery point to pick up samples and then immediately returned to the laboratory to drop off the samples.

Scenario 4 (hub and spoke). Transportation of DBS samples for VL testing in addition to EID samples using the same hub-and-spoke routes as scenario 3.

Outputs of the cost analysis included the monthly estimated costs for transport by type of sample, cost per health center by type of transport, and cost per sample transported by type of transport. In all scenarios, the transportation cost per km was found to be higher for UAVs than for the existing motorcycle system by at least 25 percent. A similar trend was observed in examining costs per DBS samples transported. However, in scenario 3, in which both modes of transportation only carried EIDs via hub-and-spoke routes, the *total* transportation costs for UASs were 18 percent less than those for motorcycles, including because of shorter routes and delivery times[21] and the resulting reduced vehicle and personnel costs (vehicle and equipment costs were 10 percent lower for the UAS than for the motorcycle system). Vehicle, equipment, and scheduled vehicle maintenance costs were much higher for UAS than for motorcycles in scenarios 1 and 2 but much less so in scenarios 3 and 4. The authors acknowledged that, in reality, the likely optimal system for transporting DBS and lab results will not be composed only of UAV or only of motorcycles, but will take advantage of the strengths of each technology to minimize both costs and transport time.

In 2019, VillageReach collaborated with NextWing, the Ministry of Health and Population of Malawi, and the MBTS to assess the cost of using drones to transport blood and oxytocin to treat PPH. In the cost modeling, an Excel-based tool was developed to estimate and compare operating costs (Matemba 2019), and two scenarios were explored: (1) round trip from the MBTS to all health facilities currently conducting transfusions, and (2) round trip to four health facilities located more than 30 km from the MBTS. The cost model showed that drones can have lower monthly estimated costs than ground vehicles because they take a more direct path, and traffic and poor road conditions do not affect them, thereby reducing vehicle and personnel costs (VillageReach 2019a). In both scenarios, the UAV costs were 51 percent lower than land cruiser costs. The cost of fuel was among the most significant cost drivers in the land cruiser scenarios (accounting for 11 percent of total costs), whereas the cost of the vehicle was a key cost driver in both the UAV and the land cruiser scenarios. On the other hand, personnel costs were higher—by a factor of 2.5—in the UAV scenarios because of the greater expertise required.

Another transport network optimization study commissioned by UNICEF Malawi in 2018 evaluated UAVs as an additional mode of transport in the existing specimen referral system. The study results were to be piloted in two rural districts (Nkhata Bay and Likoma Island), focusing on EID of HIV and VL testing; however, to improve equipment utilization and cost-effectiveness, TB specimens and additional health commodities such as essential medicines, vaccines, and emergency health commodities were also included (JSI 2018b). The study developed four network scenarios that serve all facilities in Nkhata Bay and Likoma Island Districts, either directly or indirectly. Up-front investment for the UAS was assumed to be $115,000, including capital investment in facilities, the vehicle (one drone to cover both districts), equipment, and office space. The following annual operating costs were estimated in addition to up-front investment:

Scenario 1 (motorcycle courier twice per week). Operating costs estimated at $37,600, primarily driven by operating costs of the motorcycle courier service.

Scenario 2 (motorcycle courier more than twice per week). Operating costs increase by approximately 8 percent over scenario 1, resulting in annual operating costs of approximately $40,700.

Scenario 3 (drones directly serve facilities inaccessible by land twice per week, as well as those facilities considered hard to reach). Operating costs are $35,000 more than in scenario 1.

Scenario 4 (uncommitted or idle time of the drone is leveraged to transport other commodities to health facilities). Operating costs vary according to the type of commodity transported. The incremental costs of layering additional use cases on the specimen referral system range from $3,500 for either emergency items or safe blood to $21,000 for vaccines.

Haidari et al.'s (2016) study on vaccine delivery by UAVs in the Gaza province in southern Mozambique found a net benefit for both the cost to deliver and product availability. The study—a modeled cost-effectiveness analysis using the HERMES software platform, which can generate a detailed discrete-event simulation model of any health product supply chain—found that drones can increase vaccine availability and decrease costs compared with standard care if drone use is maximized and optimized to overcome the initial investment and maintenance costs. In the UAV scenario, UAVs distributed up to 1.5 liters of packaged vaccines to health centers up to a 75 km radius from a hub that was colocated next to a supply depot. In the traditional multitiered land transport system scenario, trucks delivered vaccines from the provincial depot to district stores on a fixed monthly schedule and included a mix of truck and motorbike deliveries and pick-ups via public transportation to move vaccines from district stores to health centers on a monthly basis. The mean vehicle lifetime was assumed to be 10 years for land transport and 375,000 km for UAVs. The study found that implementing drone-based transport in the baseline scenario improved vaccine availability (96 percent versus 94 percent) and produced logistics cost savings of $0.08 per dose administered compared with a traditional multitiered land transport system. The UAVs maintained cost savings across all sensitivity analyses, ranging from $0.05 to $0.21 per dose administered. The minimum UAV payload necessary to achieve cost savings over the traditional multitiered land transport system was no more than 0.4 liters. The study's computational model showed that major drivers of cost savings from using UAVs are road speed of traditional land vehicles, the number of people needing to be vaccinated, and the distance that needs to be traveled.

In a study of several health systems in Sub-Saharan Africa, Wright et al. (2018) analyzed whether, for long-tail products (for example, antivenom and rabies postexposure prophylaxis), it is better to stock the products at the health facility level only or to distribute them on demand by UAVs by also considering inventory costs. Wright et al. (2018) express safety stock as a multiple of average demand at the health facility level, and assume that for these highly variable products, safety stock is four times the average demand (variability factor), and the inventory holding cost is 10 percent of the purchase price for both products. Finally, the cost per flight was assumed to be $25. For antivenom, the authors find that the cost of on-demand supply by UAS is about half the cost of stocking at each health facility and therefore is cost-effective. For the rabies postexposure prophylaxis use case, delivery on demand is more than six times the cost of holding stock at the health facility level because of the lower cost of the product, and therefore is not cost-effective.

For the Dodoma Region of Tanzania, Rupani (2017) assesses the potential cost impacts of introducing UAVs for the delivery of blood, rabies postexposure prophylaxis and immunoglobulin, vaccines and supporting products, and prioritized program and essential medicines. The analysis covers 157 health facilities. Across facilities at current delivery frequency by truck, safety stock is approximately six weeks of inventory. The study suggests that, beyond transport costs, costs can be reduced in other parts of the system by reducing inventory holding costs and capacity needs. This finding is especially relevant for vaccines, a very high-value product: each facility can easily hold more than $1,000 of vaccine inventory at any given time. The study suggests that smaller, higher-frequency deliveries enabled by UAVs could lead to reductions in safety stock and inventory holdings. As a result, significant savings could be realized, which could offset higher transport costs. In addition, higher-frequency deliveries would also mean less cold chain storage needed at the facility level.

The hypothetical "East Africa base case"

As part of the current study, cost-competitiveness modeling and associated sensitivity analysis were developed for a hypothetical "East Africa base case" for emergency medical goods deliveries using UAVs. The base case under the hypothetical model developed by Deloitte assumes that UAVs serve 20 destinations and 2,000 total annual cases, corresponding to what would be expected for a catchment population of 1 million. The analysis compares the cost savings under different drone vendor assumptions, namely a "high-price vendor" ($310,000 initial drone purchase price), a "low-price vendor" ($5,000), and a "base case vendor" ($15,000). The hypothetical vendors also differ in technical characteristics such as speed, set-up time, maximum range, and payload.[22] Based on input from drone vendors participating in the African Drone Forum in February 2020 in Kigali, Rwanda, the analysis also assumes infrastructure investment needs,[23] ranging in cost from $50,000 for the low-price vendor to three times that for the high-price vendor. The modeling explores two different business models—one involving UAV purchase and the other involving drone-as-a-service (that is, leasing of drones and all associated services); in the cheapest vendor case the annual price is assumed to be $60,000 but reaches $420,000 for the high-price vendor. Finally, as for ground transportation modes (the counterfactual), the base case assumes half of all cases to be served by motorcycles and the remainder by a mix of land cruisers (20 percent), vans (15 percent), and a combination of vehicles and ferries (15 percent).

In the drone purchase and the drone-as-a-service scenarios, using only the cheapest vendor is estimated to result in positive net savings of about $20,700 and $88,300 annually, respectively (figure 1.14). In the high-price vendor scenarios, net savings are highly negative, at roughly −$300,000 depending on the vendor pricing model. Net savings are also found to be very sensitive to the assumed infrastructure cost and annual cases served. For example, an increase in annual cases served from the baseline 2,000 to 18,840 (corresponding to a 10-fold increase in catchment population) would generate positive net savings of nearly $320,000. The estimated rule-of-thumb elasticity with respect to ground distance that would need to be traveled is about 0.5, meaning that a 10 percent increase in distance would be expected to increase net savings from using UAVs by 5 percent.

Overall, the sensitivity analysis demonstrates that any savings in transport costs for medical goods using UAVs as compared with traditional transport

FIGURE 1.14

Scenario analysis of annual costs and savings for a hypothetical East Africa case

a. Vendor comparison, drone purchase

	Vendor A - low price	Base case - medium price	Vendor B - high price
Net savings	$20,734 +23%	–$89,958	–$320,708 –256%
Yearly operating cost	$77,563 –43%	$138,255	$319,005 +130%
Infrastructure cost	$50,000 –50%	$100,000	$150,000 +50%
Ground transport savings	$148,297 0%	$148,297	$148,297 0%

b. Vendor comparison, drone-as-a-service

	Vendor A - low price	Base case - medium price	Vendor B - high price
Net savings	$88,297 +178%	–$31,703	–$271,703 –757%
Drone-as-a-service price (annually)	$60,000 –66%	$180,000	$420,000 +133%

Source: Modeling by Deloitte for current study.

modes are mostly driven by drone vendor pricing structures and drone costs, ground transportation costs, and the level of demand:

- *Drone vendors vary greatly in their pricing and drone capabilities.* One drone may be nearly 10 times the price of another, so to be cost competitive, the cheapest vendor that meets the required specifications should be chosen.
- *Leasing provides a much more cost-competitive option compared with buying.* Leasing not only often results in lower overall costs but also avoids large capital investment requirements and increases cost stability compared with owning drones. In the hypothetical East Africa case analysis, leasing is the only cost-competitive option across all scenarios.
- *Because drones cannot compete on price with cheap or free transportation (including public transit), site selection is critical to a cost-effective drone program.* Drones must be used where ground transportation is currently expensive.
- *Drones scale well because savings add up faster than costs.* When the population increases by a factor of 10, traditional ground transport costs increase by roughly six times, while drone costs only double. This difference allows large savings to be realized from not using ground transportation, which offsets the costs of the drones. However, having a prohibitively expensive drone or cheap or free ground transportation will prevent drones from ever reaching a cost-competitive scale, no matter how many trips the drones travel.
- *Infrastructure costs can easily prevent drone programs from becoming cost competitive, regardless of demand levels or drone vendor specifications.* Offsetting these costs by either leasing a drone or selecting the drone with the cheapest capital investment is critical to being cost competitive.

Ukerewe District, Tanzania

Health facility–level analysis conducted in Ukerewe for the current study examined several scenarios to determine the most cost-effective medical goods use cases from the perspective of substituting UAV-based transportation for the current land- and boat-based transport modes. The analysis also assessed the most cost-effective routes and the associated needs for number of flights and number of UAVs required. The study made several specific assumptions about the parameters of the UAVs that could be used to meet the needs of the medical goods use case in the specific geographic context. These assumptions include UAV cargo capacity of 12 liters or 4 kg, a range of 100 km, and an equipment life cycle of approximately 1,000–1,200 flights. The UAV (a small hybrid drone powered by alternative fuel) is assumed to have an initial capital cost of $75,000, which corresponds to a UAV with the assumed cargo capacity and range as well as VTOL capability and the ability to perform outbound and return transport. The cost includes all necessary equipment, such as fuselage and battery. In subsequent scenarios, the capital cost of the UAV—*assuming equivalent technical specifications*—was reduced to $20,000 for comparison.

The analysis assumes that the cost per UAV is incurred in year one, and that UAV variable operating costs include staff time, electricity, internet, and maintenance. The UAV is assumed to require two hours total to complete a flight, including flight time, preparatory work, and packing and unpacking of the UAV. Considering these time requirements, it is assumed that a single UAV can make up to 1,200 flights per year. Finally, given the specifics of the road network in Mwanza, a road network circuity factor of 1.6 is assumed (where a factor of 1 applies to a road that travels from the origin to the destination in a straight line).

Lifesaving items

The infrequent needs and small quantities of lifesaving medicines such as oxytocin and rabies vaccines indicate few flights (only 432 per year across all Ukerewe facilities combined), meaning that more fixed costs are borne by each flight and there is little opportunity for savings.[24] Total transport costs if using the $75,000 UAV would amount to $80,907 per year, compared with the status quo transport costs of $20,595 (see figure 1.15); thus, the cost increase would be high in percentage terms but relatively low in dollar terms. The low number of flights and small volumes also suggest that annual quantities can be delivered using only one UAV. The weight and volume capacity utilization of the UAV is less than 20 percent, given that only 0.82 kg is carried per flight, meaning that other items could be combined with the delivery of lifesaving items during the same flight (there are opportunities for use case layering).

As a comparison, if the initial capital cost of the UAV is assumed to be $20,000, total transport costs for lifesaving items, at $25,907 annually, become much more comparable to the current method.

Vaccines

The delivery of routine vaccines and associated supplies (syringes, diluents, and so on) by UAV would require considerably more resources than the current system.[25] The increase in annual transport costs of supplying the Ukerewe facilities would be high in both percentage and dollar terms, rising from $16,554 per year at present to nearly $1.74 million per year if using UAVs. This high cost is primarily explained by the large quantities involved—a volume of about 466,000 liters

FIGURE 1.15

The cost of transporting medical goods to all Ukerewe facilities using UAVs compared with current methods, assuming an initial capital cost of $75,000 per UAV

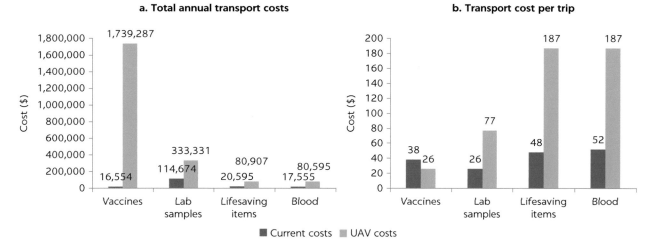

Source: Field research in Ukerewe District, Tanzania, fall/winter 2019–20.
Note: UAV = unmanned aerial vehicle.

per year to Nansio District Hospital alone. As a result, the UAS would require more than 100 flights to deliver the same quantity as land transport, given that the UAV can deliver no more than 9.3 liters per flight. Across all Ukerewe facilities combined, 65,988 flights per year would be required to meet routine vaccine demand. The high number of flights drives down the average cost per flight, but the savings would be vastly outweighed by the number of UAVs required.

Under a hypothetical assumption that a UAV with the appropriate technical specifications could carry 10 kg instead of the currently assumed 4 kg, delivery requirements would decrease from 66,000 flights to 51,500 flights per year (still requiring 40 or more UAVs to serve all the facilities), and total transport costs would decrease from $1.74 million to $1.35 million.

The UAV costs for vaccine deliveries include neither the initial capital investment for purchasing the UAV nor depreciation to enable direct comparison with the current system. Therefore, even if the initial capital cost for the technically equivalent UAV were much lower ($20,000), total estimated transport costs for delivering vaccines by UAV would remain as high as in the $75,000/UAV scenario.

Blood for transfusion

Delivery of blood and samples by UAV would require more flights than the number of trips made under the current system for some facilities because of cargo capacity constraints; for example, for the small island facilities, 192 flights per year would be required, compared with 96 trips using current transport methods. The additional flights drive up the costs. On the other hand, the small quantities delivered mean that all deliveries could be made using one UAV, making total costs relatively low compared with other medical goods categories. Similarly to lifesaving items, despite a large percentage increase in total annual transport costs compared with current methods, the use of UAVs for blood deliveries would lead to a relatively modest absolute cost increase: the cost would be $80,595 for all Ukerewe facilities combined, compared with the status quo of $17,555.[26]

As with lifesaving items, if the UAV capital cost were to be significantly lower ($20,000), the annual transport costs of blood by UAVs, at about $25,595, would become much more comparable to the status quo.

Laboratory samples

To meet the overall lab sample transportation needs for Ukerewe, 4,356 flights per year would be needed, requiring more than just one UAV. UAV-based transport would imply a significant absolute increase (although a more modest percentage increase than for other medical goods, such as vaccines) in annual transportation costs compared with the status quo: $333,331 compared with $114,674. However, lab samples may be an example of where UAV-based deliveries could be cost-effective for some types of facilities or routes. Destinations with regular flights (one or two per week) and longer distances, such as between Nansio District Hospital and Mwanza and from Nansio District Hospital to the small islands, would result in UAV costs that are the most similar to the current transport system of any medical goods category considered. For Nansio District Hospital, UAV-based transport would be associated with an overall annual transport cost of $16,771, compared with the current transport cost of $15,402; for the small islands, the respective costs would be $37,744 for UAV transport and $34,660 under current methods.[27]

At $75,000 per UAV, the total cost of delivering laboratory samples in Ukerewe is three times that of the status quo, but at $20,000 per UAV the costs become roughly equivalent.

Cost dependency on the context: Number of flights, opportunities for case layering

Previous cost-effectiveness analyses, such as Wright et al. (2018), compared various transport options for a variety of delivery categories using UAVs vs. well-managed traditional modes of last-mile delivery. Overall, their findings show that UAV cost-effectiveness is driven primarily by the number of flights per year that can defray fixed costs, and that increasing flight numbers is dependent on facility density within the UAV range area. Flight numbers can be increased by operating in areas with higher health facility density and selecting UAVs that have longer ranges or by layering multiple use cases.

The number of facilities in range of a UAV is driven by two factors: the range of the UAV and facility density in the region. The range of a small fixed-wing UAV is about 7.5 times that of a multicopter, meaning that the number of facilities falling within the applicable delivery area would differ by a factor of more than 50. In Mwanza, Tanzania, 253 facilities would be within range of a small fixed-wing UAV, compared with about 180 for a small hybrid, about 30 for a medium hybrid, and just 5 for a multicopter. For the Nkhata Bay area in the Northern Region of Malawi, JSI (2018c) estimates that, assuming the base of UAV operations is at Nkhata Bay District Health Office, 9 facilities would be in the range of a small hybrid UAV (range 60 km), 15 facilities in the range of a medium hybrid UAV (100 km), only 1 facility within the range of a multicopter (20 km), and 20 facilities (or all facilities in the Nkhata Bay and Likoma Island districts) within the range of a small hybrid gas/electric UAV or a small or large fixed-wing UAV.[28]

According to the cost curves developed by Wright et al. (2018), small fixed-wing UAVs are the only ones that drop to a lower cost per km than motorcycles in a well-managed fleet (with a transport cost of $0.30–$0.40 per km for the motorcycles), and they do so only after about 5,000 flights. The cost and performance advantage of UAVs is enhanced when the road circuity factor is high, the motorcycle operating cost is high, and many facilities are in range. The most suitable and cost-efficient option would be a small, fixed-wing UAV that has a relatively long range and that would not need to land to pick up cargo at the delivery site.

Based on the Ukerewe context, and taking lifesaving items as an example of the product to be delivered, the JSI Tool for Determining Cost Effective Use Cases for UAVs suggests that, for example, using a multicopter UAV, the cost per km would be equivalent to that of a land cruiser at about 7,500 flights per year; however, for a small hybrid drone powered by alternative fuel, no foreseeable number of flights would allow the UAV cost per km to decline sufficiently to be at parity with the land cruiser mode (see figure 1.16). The cost-effectiveness of UAVs as compared with land transport modes in a different context (Ghana) was analyzed by Drones for Development (2016). Their analysis suggests that, although the expected return on investment in the UAVs would take only about a year, the size of the required up-front investment to serve just a single district of the country is substantial (see box 1.4).

For regular medical supplies (as opposed to emergency goods), stock-outs might provide a good example of where value is achieved through product layering. Wright et al.'s (2018) findings demonstrate that the single-use-case approach to introducing UASs is not optimal from a cost-effectiveness perspective. Except for safe blood for transfusion, their analysis shows that using UAS for single-product-category deliveries is not optimal, and that layering multiple use cases will increase UAS cost-effectiveness by increasing the number of flights the UAV will be used for. The authors suggest that, based on the number of flights per day, a possible use case would be to combine vaccine delivery with diagnostic samples transportation, with the possible inclusion of vertical program commodities (HIV medicines and TB and leprosy medicines).

"Layering" several medical use cases offers the potential to reduce average costs per flight by leveraging the existing uncommitted time of a UAV.

FIGURE 1.16

Transport cost by UAV or land cruiser vs. anticipated number of flights per year

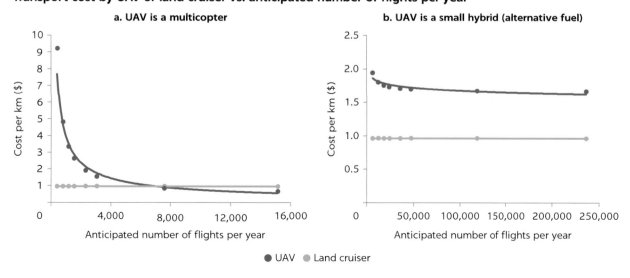

Source: JSI Tool for Determining Cost Effective Use Cases for Autonomous Aerial Systems, estimates based on Ukerewe District, Tanzania, context.
Note: km = kilometer; UAV = unmanned aerial vehicle.

The Dr.One business case (Ghana)

The Dr.One concept is focused on "last-mile delivery" (defined here as last 50–100 km) of health care commodities such as drugs, contraceptives, blood samples, rapid diagnostic test kits, and vaccines. Dr.One is a scalable design using low-cost components to transport up to 2 kg. The business case analysis for Dr.One concluded that the deployment of drones for the transport of health commodities to remote locations to augment the existing motorbike supply chain system is cost-effective. On the basis of actual health care data, the need for 500 Dr.One flights per year in the Builsa South District alone could be established. Actual cost savings for the Ghana Health Services were estimated at $4,129 per drone per year if Dr.One drones were to be introduced in small quantities, for example, at one location in the Builsa District. The main cost savings can be attributed to differences in fuel costs and vehicle speed. These two factors alone account for most cost

savings. The fact that drones fly in a straight line whereas motorbikes follow a longer route via the road structure accounts for additional but smaller cost savings. Savings on wear and tear of motorbikes, replacement of parts, and accidental damage to motorbikes are smaller and hard to predict. Recruitment and training of qualified personnel for the launch and recovery of drones adds to the costs. Builsa District is assumed to be representative of all other districts for use case scenarios. Linear scaling of the assumed savings of $4,129 per drone per year results in total savings of $412,900 for all of Ghana. Initial procurement costs for the 100 drones ($5,000 each, including rechargeable battery packs) are $500,000. Hence, the expected return on investment in the drones would take slightly more than one year. The resulting 50,000 flights per year, saving $8 each, would serve a population of 20 million with an initial investment of only $0.025 per person.

Source: Drones for Development 2016.

This approach is confirmed by the research conducted for the current study in Ukerewe District in Tanzania (assuming that 1,200–1,600 flights per year can be conducted by the same UAV given the preparation, packaging, and flight-related time requirements). The analysis focuses on two different scenarios for use case layering: one using lab sample transport as the base case and the other using blood transport as the base case.

Base case: Laboratory samples

The base case includes only deliveries of lab samples for Nansio District Hospital and the small island facilities (together amounting to 612 required flights per year), given that the annual flights required to service the big island facilities are by themselves sufficient for fully utilizing several UAVs. The 612 flights per year translate into time utilization of just 51 percent of the assumed theoretical maximum for a given UAV. As a result, the fixed costs that must be absorbed by each flight are high, leading to an overall average cost per flight of $143 (figure 1.17).

The uncommitted time can be further utilized if the same UAV is also used to transport lifesaving items to the big island facilities that are located farther than 10 km from Nansio District Hospital and to the small island facilities. Doing this would increase annual time utilization of the UAV to 74 percent and lower the average per-trip transport cost to $102. Finally, to increase the UAV's utilization even further, it could be used to transport rabies vaccines to all health facilities in Ukerewe District. As a result, the UAV would be 80 percent utilized, and the overall average transport cost per trip would be $97. The total annual transport cost

FIGURE 1.17

Layering medical goods deliveries: Effect on annual number of UAV flights and average transport costs per flight

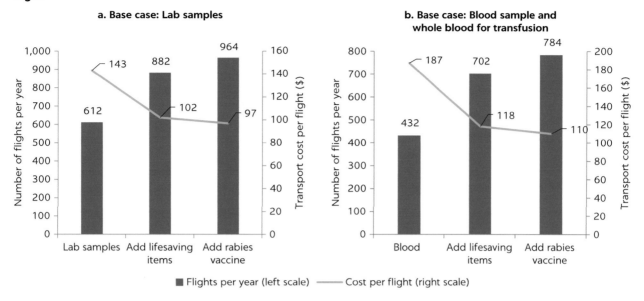

Source: Field research in Ukerewe District, Tanzania, fall/winter 2019–20.
Note: UAV = unmanned aerial vehicle.

would slightly increase from $87,366 if the UAV were to only transport lab samples to $93,122 if it were also used to carry lifesaving items and rabies vaccines.

Base case: Blood samples and whole blood for transfusion

In this scenario, lifesaving items for the small island facilities and the big island facilities located more than 10 km from Nansio District Hospital and, subsequently, rabies vaccines for all Ukerewe health facilities, are layered on top of the base case consisting of blood sample and whole blood deliveries for all Ukerewe health facilities. Layering other medical goods on blood makes economic sense because blood delivery for all Ukerewe facilities could be completed with a single UAV and still leave 64 percent of its time uncommitted. The results are similar to the lab samples base case: the total flights made by a UAV can be increased from 432 to 784, with an associated increase in the UAV's time utilization from 36 percent to 65 percent and a decline in average transport costs from $187 per trip to $110 per trip.

In summary, as a result of case layering, the time utilization of a given UAV could be increased by 60–80 percent, leading to average transport cost savings per flight of about 30–40 percent.

In principle, it is also possible to take advantage of the excess physical capacity of a given UAV, such as for routine items, given that the shipments can be scheduled and coordinated. However, layering emergency-type items on top of routine items or lab samples is less feasible, given that emergencies are random, unplanned events. If a routine item was about to be sent off at the time that an emergency need arose, the emergency item could be included with that delivery, but it would need to be going directly to that one facility at the hour needed.

The relevance of the number of flights and the importance of use case layering in the Malawi context is also demonstrated by analysis conducted for the current study based on an Excel-based simulation tool developed by Deloitte.

Using health facility–level data provided to the World Bank by the Malawi Ministry of Health and supplementary data provided by other development partners, the analysis suggests that any savings in transport costs of medical goods using UAVs as compared with traditional transport modes are driven by (in order of relevance) level of demand, mode of traditional transport, and difficulty in reaching the health facilities. Layering use cases is critical to maximizing UAVs' operational effectiveness: if they make only one or two trips a week, becoming cost competitive would be nearly impossible.

UAV cost parity with current transport options

The cost parity of UAVs with traditional transport modes in East Africa and in Sub-Saharan Africa more broadly has been analyzed by Wright et al. (2018). The authors suggest that the current cost per km for a motorcycle in a well-managed fleet in a rural African public health setting is approximately $0.33; however, for any given trip a motorcycle will travel a longer distance by road and the UAV will fly in a straight line. Assuming a road circuity factor of 1.5, the cost per km is $0.50. The authors estimate the cost per km for a small hybrid UAV with more than 5,000 flights annually from a single hub to be approximately $1. Given these figures, the authors estimate that it will take seven years for small hybrid UAVs to become transport-cost competitive with motorcycles (on cost per km) assuming 10 percent improvement per year, three years assuming 20 percent improvement per year, and two years assuming 30 percent improvement per year. However, if the UAV range is greater than assumed (60 km), the costs are lower than currently assumed, or the geography is particularly inaccessible by road (higher road circuity factor), it is possible that small hybrids may already be transport-cost competitive with motorcycles (on cost per km). The authors analyze routine vaccine resupply systems in Sub-Saharan Africa based on the model of monthly direct deliveries to health facilities by multistop motorcycle transport and compare that with more frequent delivery by a UAV or motorcycle. For cost alone, they find that UAVs are not competitive with a motorcycle with a multistop route, even if motorcycle delivery frequency is weekly, suggesting that UAV costs would have to be reduced by half or more to be cost competitive with a multistop route by motorcycle.

Ukerewe District, Tanzania

Field research undertaken in Ukerewe for this study estimated the UAV-associated costs that would bring this technology to cost parity with currently used transport modes. It should be noted again that, in this specific context, public transport—an inexpensive mode—is widely used for transporting public health commodities. Using existing transport modes, the operating cost per liter delivered currently ranges from $0.58 for routine vaccine deliveries from the mainland (Mwanza) to Nansio District Hospital to $61.71 for lab sample pickup from the small islands. To express the UAV-relevant cost parity in dollars per liter per km terms, the current operating costs per liter must be divided by the "as the crow flies" distance to each type of destination, by product delivered, considering the commodity flow patterns illustrated in the section titled "Present costs and modalities." For example, the straight-line distance for vaccine deliveries from Mwanza to Nansio District Hospital is approximately 50 km, but is about 90 km for delivery of lifesaving items to—and lab sample pickup from—the small island facilities (also serviced directly from the MSD in Mwanza).

The estimated UAV operating cost parity is extremely low for vaccine deliveries: regardless of the route type, parity is less than $0.01 per liter per km (figure 1.18). For blood deliveries, the operating cost parity is more variable depending on the route type, ranging from about $0.10 to the big island and the small islands to about $0.35 per liter per km to Nansio District Hospital. For lifesaving items delivery, estimated parity would be as high as $0.55 per liter per km for deliveries to the small islands but less than $0.02 per liter per km for deliveries to the big island facilities located within 10 km of Nansio. Finally, the estimated parity is by far the highest, meaning that UAV-based transport would have the best chance at being cost competitive with overland transport, for lab sample pickup from the small island facilities and the big island facilities located farther than 10 km from Nansio, at about $0.68 per liter per km.

The UAV capital cost that would allow this technology to break even with current transportation modes used in medical goods deliveries in Ukerewe is estimated to be about $19,000. This estimate is significantly less than the $75,000 capital cost per UAV assumed throughout the cost analysis but not impossible to reach given the speed with which the technology is developing, including in research and development facilities in emerging markets such as China. However, it should be noted that the estimated capital cost parity assumes several specific UAV technical capabilities, such as vertical takeoff, cargo capacity of at least 12 liters or 4 kg, and a range of 100 km, which the cheaper UAV would also have to meet. The current phase of the Deliver Future / Drone X project in Mwanza intends to test new drone types and to extend the network beyond Ukerewe to the neighboring islands. It also aims to establish a concrete business case for low-cost UAVs and to evaluate potential drone delivery markets and the potential for drone initiatives to create jobs for local inhabitants (Rabien 2018).

FIGURE 1.18

Operating cost parity for UAVs, considering current transport costs

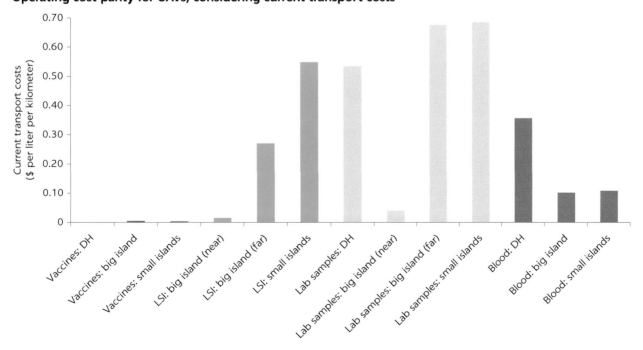

Source: Field research in Ukerewe District, Tanzania, fall/winter 2019–20.
Note: DH = Nansio District Hospital; LSI = lifesaving items; UAV = unmanned aerial vehicle.

FIGURE 1.19

Cost parity for a hypothetical use case in Malawi: Break-even price at current scale

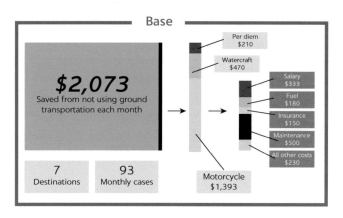

Source: Analysis by Deloitte conducted for the current study.

Malawi

Malawi-focused cost modeling conducted by Deloitte to provide complementary insights to the hypothetical East Africa base case examines scenarios assuming a drone-as-a-service model. It finds that cost parity—or maximum cost-competitive monthly all-in drone-as-a-service cost—is approximately $2,000, or equivalent to the savings obtained by not using ground transportation (figure 1.19). The case study also highlights the importance of the level of demand in determining the maximum price at which UAV services would be cost competitive with ground transportation. That is, if the number of monthly cases were to quadruple, the parity monthly drone-as-a-service price would more than double, to $5,000.

PUBLIC HEALTH IMPACTS

It is clear that cost-benefit assessments of UAVs in medical goods delivery should consider not only monetary cost impacts but also health benefits. One of the challenges for establishing the long-term rationale for drone applications in medical goods deliveries is the lack of standard monitoring and evaluation frameworks in the region's countries for measuring health and supply chain impacts (Machagge 2018). Incorporating health benefits in cost-benefit assessments requires the availability of baseline data on desired health outcomes and an estimation of interventions *not* provided because of limitations in physical access to commodities. Evaluating the cost-effectiveness of drone operations is directly linked to the ability to assess the resulting health impacts, given that it is a relative measure that considers the cost required to achieve a specific outcome, such as dollars per DALY or dollars per life saved (USAID 2018).

Because of the early stage of implementation and limited small-scale pilot projects in most Sub-Saharan African countries where UAVs have been used for medical goods deliveries, limited primary data are available related to the health impacts of UAVs. One of the main recommendations for governments, implementers, drone providers, and funders highlighted in Knoblauch et al. (2019) is the need to evaluate the impacts using standardized indicators. Given that drone-specific data are not recorded in routine health or laboratory information systems, the authors also propose that a "drone information system" be created to cover all drone-related flight log and telemetry data, as well as the direct integration of certain drone information system data such as "dispatched and received" logs about payload into the existing stock registries in health facilities.

Although one of the main effects expected as a result of integrating UAVs into health supply chains is an increase in the number of people with access to targeted medical products and services, the feasibility of evaluating the actual health impacts on the target population is higher for some medical conditions than others. For example, if UAVs are used to transport vaccines over a period of 12 months, the targeted health outcomes of reduced incidence of vaccine-preventable diseases may not be observed within the 12-month period even if it may be possible to estimate the consumption of the vaccines provided to the sites served (VillageReach 2019b).

This section aims to contribute evidence about the public health impacts that can be expected as a result of introducing UASs in specific medical goods deliveries and route typologies, based on field research conducted for this report at the health facility level in Ukerewe District, Tanzania.

Improving survival rates and treatment effectiveness

Logistical constraints in medical supply chains can cause a temporary shortage of supply; in turn, those delays can have a significant impact on the patient, potentially resulting in a higher mortality rate. In most cases, a delay can be managed by using other medical options (for example, oxytocin in PPH cases in which blood for transfusion is not readily available). However, there are no alternatives, for instance, if HIV or TB drugs or insulin are missing (Dirks 2017). For pediatric HIV, early identification of children who were born with HIV is crucial as a first step in securing their treatment and care. Logistics is one of the crucial bottlenecks when every day counts (UNICEF Malawi 2019). The longer the delay between the test and the results, the longer the patient must wait for treatment, worsening the child's health status (USAID 2017). Turnaround time is also crucial for certain other lab sample tests, especially for diseases that are highly contagious, as has recently been visibly demonstrated in the context of the COVID-19 pandemic, for which rapid test results can not only save lives but also provide a more viable path for the economic reopening of communities.

Faster and smoother than road travel, and offering greater thermal control, drones could ensure that samples arrive intact, and the impact of rapid and accurate results could potentially be significant. In emergency situations, such as complicated labor requiring transfusion, the timely arrival of safe blood for transfusions could determine a woman's survival. The prompt use of blood products, including packed red blood cells, plasma, and platelets, can also save the lives of bleeding trauma patients (Thiels et al. 2015). In all cases of blood transfusion, minutes can be essential. However, for other chronic diseases, the difference of a day or two may not matter as much, for example, VL monitoring, if the specimen is properly separated and stored.

In the context of East Africa and Sub-Saharan Africa more broadly, the other bottlenecks beyond timely availability of drugs and supplies that may prevent health impacts from materializing must be considered. For example, in Tanzania, although the share of health facility births (deliveries) according to UNICEF has increased from 47 percent in 2005 to 63 percent in 2015, the coverage of basic emergency obstetric and newborn care remains low, with only 20 percent of dispensaries and 39 percent of health centers offering delivery services that provide all signal functions. There are wide gaps in births assisted by skilled health professionals in rural and urban areas, assessed at 55 percent and 87 percent, respectively.

Data received from a sample of seven hospitals across Malawi illustrate the human health impact of current gaps in the emergency and essential medicines supply chains discussed in earlier sections of this chapter. About 6 percent, on average, of all PPH patients in these facilities died during the period between July and September 2019, partly as a result of delays in administering blood transfusions; for snakebite, 4 deaths were recorded out of 37 total cases. Among the 148 total cases of stroke, 10 resulted in death and another 49 in long-term disability.

Since its inception, Rwanda's UAV-based blood delivery project has provided emergency blood supplies to more than 1,200 women, greatly reducing the risk of maternal mortality, and reduced wastage of blood by 4 percent in the logistics network that is serviced by Zipline (WEF 2018). In 2019, the company reported that its services had helped save more than 1,500 lives, given that approximately one-third of the blood deliveries are for emergency replenishments, mostly for mothers and children: about 50 percent of blood transfusions benefit mothers experiencing PPH and 30 percent benefit children with severe anemia.

Pretransfusion blood samples and whole blood for emergency transfusion were prioritized by JSI (2018a) for drone-based deliveries in Tanzania based on public health importance. Blood samples for testing suffer from hemolysis due to poor transportation, and the emergency blood for rare blood groups in particular is vital to saving the lives of women giving birth who experience excessive blood loss. The use of UAVs could ensure quicker delivery of supplies and that clinicians spend more time at facilities treating patients. However, for chronic diseases affecting the Mwanza population, the speed and responsiveness of UAVs was not considered likely to offer a sufficiently compelling advantage over motorcycles to outweigh cost considerations (Wright et al. 2018).

In the USAID GHSC-PSM in Malawi, the focus from the beginning has been on reducing HIV rates and improving HIV treatment rather than on cutting the costs of transporting lab samples. Before the project-funded drone activity began, local hospital staff were absent from their health facilities for at least a week each time they dropped off lab samples and obtained sample results, which took them away from their primary care provision roles (Aryal and Dubin 2019).

The second pilot phase of the Deliver Future / Drone X project in Mwanza, in implementation since mid-2018, is specifically focusing on providing a better understanding of the health impacts of the drone corridor—through data mapping and evaluation over 18 months—as well as on knowledge transfer through the training of 12 operators and 6 technicians and engineers, with the goal of transferring the Mwanza-Nansio delivery route to the local team after six to eight months (Rabien 2018).

Ukerewe District, Tanzania

The hypothetical health impacts, that is, deaths averted, of UAV-based deliveries of blood for transfusion in Ukerewe District were assessed as part of the current study, based on collected health facility–level data and interviews with local health workers to establish the time sensitivity of blood delivery for transfusion from the perspective of patient survival rates. Data for this part of the analysis were collected from the only four health facilities where blood transfusions are performed in Ukerewe: Nansio District Hospital and three other health centers: Muriti and Kagunguli on Ukerewe Island and Bwisya on Ukara Island. The health impact was estimated using a decision-tree model, which represents the sequence of events that occur when a patient that needs an emergency blood transfusion arrives at a health facility; the model is generic enough to also be used to estimate the health impact related to other emergency products.

The analysis determined the unmet need for blood transfusion (emergency orders requiring blood) in the four facilities based on current data; to do that, indicators such as the number of cases requiring blood transfusion and the amount of blood required, the amount of blood collected each week, and the amount of usable blood available were considered. The analysis also compared the time required to deliver the ordered blood using current transport modes and UAVs if blood were not available when needed. Patient survival rates were assessed using the decision-tree model relating the likelihood of survival to (1) timely versus delayed blood delivery (either from Mwanza or from the nearby facilities) and (2) the quality of the blood delivered (in good condition or poor quality). In estimating the survival rate impacts, the analysis defined a "late delivery" as a delivery that takes more than four hours; it also assumed that a late delivery would most likely lead to a fatal outcome, a case fatality rate of 98 percent. In addition, it was assumed that traditional transport reduces the quality of the product more often than UAV transport would. Finally, based on Drammeh et al. (2018), the analysis assumed a baseline case fatality rate of 5.2 percent for patients receiving blood transfusion in Tanzania.

Across Ukerewe District, about 100 units of blood, corresponding to 55 patient cases, are required each week, compared with the blood availability of 90 units (figure 1.20). The unmet need for blood is thus estimated at 10 units per week, equivalent to five cases per week or 260 units per year. Most of the blood required, 71 percent, is for pregnant women and children younger than age five; about 40 percent of the need (75 cases per year) is for emergencies. For these cases, about 5 units each week can be received as a result of exchanges from the other Ukerewe facilities; thus, the probability of not being able to cover the 10 units of unmet need for blood each week is one-half.

At baseline, the analysis assumes that there is a 95 percent probability of timely delivery using UAVs compared with a 5 percent probability by traditional transport, given that the ferry takes four to six hours and travels once per day.

FIGURE 1.20

Blood transfusion needs vs. blood availability in Ukerewe District

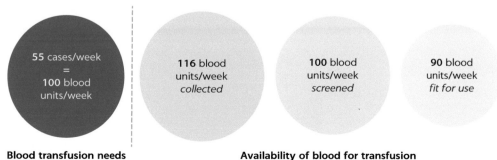

55 cases/week
=
100 blood units/week

116 blood units/week *collected*

100 blood units/week *screened*

90 blood units/week *fit for use*

Blood transfusion needs   **Availability of blood for transfusion**

Source: Field research in Ukerewe District, Tanzania, fall/winter 2019–20.

Therefore, at baseline, it is very unlikely that blood will be transported from Mwanza to the facility within the critical four-hour window using current transport modes.

The main result of the analysis is that the deployment of UAVs to deliver emergency blood for transfusion to health care facilities in Ukerewe would reduce the number of deaths by almost 80 percent, from 40 deaths per year to 9, compared with traditional transport (figure 1.21), by enhancing timely blood deliveries from nearby transfusing facilities and from NBTS in Mwanza.

Sensitivity analysis. A number of assumptions about the demand for blood for transfusion, the critical time threshold for the transfusion, and other factors affect the estimated savings that could be achieved by introducing UAVs into the blood-for-transfusion transport supply chain in Ukerewe. Key findings from the sensitivity analysis conducted to quantify the degree of uncertainty, depending on the assumptions, are summarized below.

- The hypothetical health impact of deploying UAVs for blood delivery in Ukerewe would increase with an increase in the number of people with unmet need for blood for transfusion. At the baseline of 75 emergency cases per year with an unmet need for blood, the use of UAVs would help avert 31 deaths (reducing them from 40 to 9); however, if the number of emergency cases with unmet need for blood were to increase to 150, the impact of the UAVs would translate into 63 deaths averted (81 deaths with traditional transport vs. 18 with drones) (figure 1.22).
- On the other hand, the impact of using UAVs compared with traditional transport would be lower if the probability of obtaining blood from other Ukerewe facilities (as opposed to having to order blood from Mwanza) were higher than it currently is. Applying the same logic, the expected health impact of UAVs would increase if the probability of obtaining blood from nearby facilities were lower than at present.
- The health impact of UAVs would, of course, be lower than currently estimated if the timeliness of deliveries by traditional transport (ferries) were to significantly improve compared with the current journey of four to six hours

FIGURE 1.21

Ukerewe District: Estimated number of deaths per year with traditional transport vs. UAVs as a result of extreme blood loss

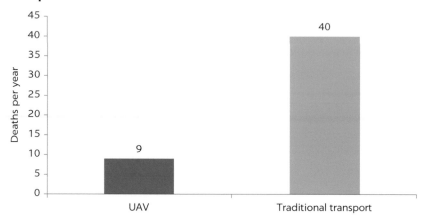

Source: Field research in Ukerewe District, Tanzania, fall/winter 2019–20.
Note: UAV = unmanned aerial vehicle.

FIGURE 1.22

Sensitivity of UAV health impacts (deaths averted) to number of cases and probability of timely blood delivery by traditional transport

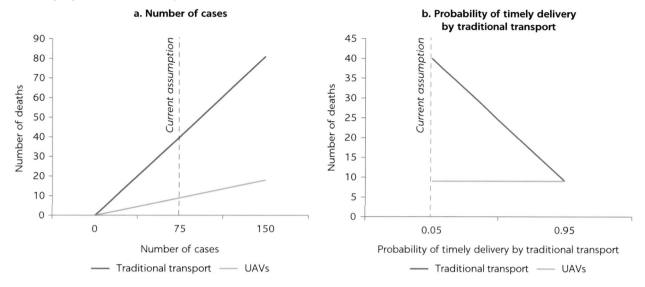

Source: Field research in Ukerewe District, Tanzania, fall/winter 2019–20.
Note: UAV = unmanned aerial vehicle.

once a day that makes delivery of blood within the critical time window nearly impossible if blood must be shipped from Mwanza. However, UAVs would still provide health benefits as measured by lives saved if the probability of timely delivery by traditional transport remains below 95 percent (equivalent to the assumed probability of a timely delivery by drone); this is nearly a given in the Ukerewe context.

- Finally, the health impact of UAVs would be reduced if a delay in blood transfusion of more than four hours is less likely to lead to a fatal outcome than currently assumed (the baseline assumption being that a delay of more than four hours results in death in 98 percent of cases). A decrease in this probability, which is unlikely, would reduce deaths more in the traditional transport scenario than in the UAV scenario. However, even if this probability is assumed to be as low as 25 percent, UAVs would still help save 10 to 11 lives per year.

It should be noted that in the Tanzania context, which likely resembles the situation in the region more broadly, public transport is inexpensive and is widely used for transporting public health commodities. However, schedules can be limited and generally inflexible, constraining the health centers' ability to respond as needed. The facility-level data analysis confirms that health care workers are often tasked with the delivery and collection of goods, which takes up a part of their time. The present analysis includes the cost of salaries and per diem in transport costs where applicable; however, it does not explicitly quantify the opportunity cost, that is, the inability to serve patients, of this facility staff time spent on picking up and delivering medical goods. Although the associated impact may be negligible for facilities near the collection points, for distant facilities or for those with limited personnel the travel time requirements can be significant, effectively replacing time serving patients, which may lead to impacts on patients' health over the long term that may or may not be possible to quantify.

Serving previously unserved demand

The implementation of UAV-based deliveries might not always translate into immediately identifiable health outcome changes but may be noticeable through surrogate endpoints, such as shorter delays in lab sample referrals or a reduction in medical supply stock-outs (Pai, Schumacher, and Abimbola 2018). Across program and essential medicines, and across geographies in Sub-Saharan Africa, stock-outs are common and large, with 20–30 percent of facilities experiencing a stock-out of at least one-month duration in any quarter (Wright et al. 2018). In 2016, 12,500 stock-outs occurred in the Mwanza region of Tanzania, compromising health services. A share of the stock-outs are caused by central shortages, whereas others are due to facility-level issues, such as incorrect ordering or forecasting of demand, or variability in lead time for distribution or accessibility in the wet season.

For multidrug-resistant TB, more accessible local treatment, which can range from 6 to 24 months depending on the type of resistance, could reduce barriers to continued treatment and reveal hidden demand in the Mwanza region (Wright et al. 2018).

On-demand response models that include UAVs may also result in increased availability for sporadic-demand products. For example, as shown by a study focused on Dodoma Region, Tanzania, the region's health facilities often do not hold rabies postexposure prophylaxis in stock because it is an unpredictable, infrequent-use medicine with less than one bite per month observed per facility (about 1,700 cases per year across all facilities). Tracking inventory and ordering for such a sporadic-demand product is a low priority. As a result, many patients must make multiple trips to multiple facilities to obtain treatment, at high personal cost, and they often obtain very delayed treatment or no treatment as a result. Using UAVs, rabies postexposure prophylaxis could be provided to the health facilities on an as-needed basis, providing reliable treatment for 1,700 patients each year (Rupani 2017).

In Rwanda, the introduction of drone deliveries has made the system more responsive to demand, preventing stock-outs at local facilities while at the same time minimizing waste, a particularly important impact for blood, which has a short shelf life and must be kept refrigerated (Rosen 2017). In 2019, Zipline reported a 0 percent stock-out rate for blood products and other essential medicines at the facilities it services (down from 88.5 percent initially), as well as a 0 percent share of expired products at the facility level resulting from reduced on-site storage and centralized prioritization of near-expiring products (down from about 10 percent). As a result of reduced spoilage, hospitals have reduced their on-hand blood inventory by up to 25 percent, and health workers are able to spend more time on care provision instead of procurement. Blood component usage in the serviced facilities has increased by 172 percent compared with service levels before drones. As of early 2019, the Zipline service had increased access to rare blood products by 175 percent (McCall 2019).

The Kasungu District project in Malawi has focused on gaining an understanding of the potential human impact of drone-based deliveries, in particular, reduction of stock-outs of various medical commodities. However, the comparison with traditional methods has been difficult because of a lack of comparative data. It has therefore been challenging to establish whether truly transformational impacts could be achieved as they were in other regions where similar projects have been implemented with UNICEF assistance, such as the Pacific islands.

The introduction of drones within the overall medical supply chain can also expand the populations served by improving the up-front affordability to households of an essential health service. For example, although health policy in the Democratic Republic of Congo advocates for free vaccination services for children at all vaccination sites, the transport of vaccines overland from the Province Health Divisions to the Health Zones and from the Health Zones to the vaccination sites, as well as the cost of oil for the storage of vaccines at the peripheral level, present costs that are not taken into account by the government. As a result, some health centers charge fees for children's access to the vaccination service, ranging from $0.50 to $2.00 per child per vaccination session. This cost hinders access for many children (Gavi 2014).

Providing access to diagnosis

Finally, another important health impact of the application of UASs may be in increasing the share of the population that is aware of a medical condition they have through more timely testing and diagnosis. Education leads to empowerment and, as patients learn of their medical status, they gain the necessary information to act in support of their health. This impact applies to a variety of medical conditions common in the region, including HIV, TB, and others. For example, nearly one-third of all blood tests in 2018 were spoiled in transit from the islands in Lake Victoria to Mwanza using traditional road- and ferry-based transport (JSI 2018a), meaning that no disease screening could be performed for high-risk patients, depriving them of timely diagnosis. UAV transport can help minimize these risks.

For the DrOTS pilot in Madagascar, the primary identified benefit of introducing drones in the transport supply chain, according to the Stony Brook University staff overseeing the program, was a *doubling* in the frequency of TB diagnosis (that is, the local population's access to diagnosis). People who were suspected of having TB were now able to get tested and subsequently become part of the drone-based treatment and follow-up program, increasing their chances of recovery.

NOTES

1. "Reproductive Maternal, Newborn, Child and Adolescent Health (RMNCAH)," UNICEF, https://www.unicef.org/uganda/what-we-do/rmncah.
2. East Africa comprises Burundi, the Comoros, Djibouti, Eritrea, Ethiopia, Kenya, Madagascar, Malawi, Mozambique, Rwanda, Somalia, South Sudan, Tanzania, Uganda, and Zambia.
3. Global Burden of Disease (2017), https://vizhub.healthdata.org/gbd-compare/.
4. Global Burden of Disease (2017), https://vizhub.healthdata.org/gbd-compare/.
5. In Rwanda, 75 percent of people living with HIV receive antiretroviral drugs, and almost 95 percent of pregnant women living with HIV take antiretroviral drugs to prevent transmitting HIV to their children. As a result, mother-to-child transmission of HIV since 2016 is just 2 percent. Still, about 12,000 children younger than 15 were living with HIV in 2018 (Rwanda—Health, Humanitarian Data Exchange, https://data.humdata.org/dataset /world-bank-health-indicators-for-rwanda).
6. "Maternal and Child Health," UNICEF, https://www.unicef.org/tanzania/what-we-do /health.
7. The 90-90-90 strategy set the target to diagnose 90 percent of all HIV-positive persons, provide antiretroviral therapy for 90 percent of those diagnosed, and achieve viral suppression for 90 percent of those treated by 2020.

8. Examples of the medical uses of these emergency items include oxytocin for PPH, magnesium sulfate for eclampsia and preeclampsia during childbirth, artesunate injection for severe malaria, and IV fluids for fluid replacement, such as in cases of cholera outbreak.

9. Exchange rate $1 = 2,300 Tanzania shillings.

10. The term "unmanned aircraft system," or UAS, is sometimes used interchangeably with "unmanned aerial vehicle" (UAV); however, a UAS includes not only UAVs but also a ground-based controller and a system of communications between the UAVs and the ground-based controller.

11. The Wingcopter UAV carries a payload of up to 6 kg, reaching a top speed of 130 km/h. It takes off and lands like a helicopter.

12. Since then, a dedicated flight corridor for testing various drone applications has also been launched in Sierra Leone in West Africa, in a partnership between the national government and UNICEF.

13. Earlier studies of drone-transported biologics (for example, Amukele et al. 2016) that had been flown for about 30 minutes, equivalent to a 20- to 25-km distance, in small fixed-wing drones, at ambient temperatures between 3.3°C and 8.8°C, found that the drone transport system tested had no adverse impact on the times to growth or the other phenotypes of the sample types or microbes that were tested.

14. "What Should You Deliver by Autonomous Aerial Systems? Tool for Determining Cost-Effective Use Cases for AAVs," https://www.updwg.org/wp-content/uploads/2019/04/UAS-Cost-effective-use-cases-simulation-Tool-v3.xlsx.

15. "UAV Delivery Decision Tool," https://fhi360.shinyapps.io/UAVDeliveryDecisionTool/.

16. The estimates assume two possible drone types: a $3,000 drone that flies at 40 km/h, can carry 1 kg, and can fly for 8 hours; and a $50,000 drone that flies at 100 km/h, can carry 2 kg, and flies for 8 hours.

17. "Our Product," Micromek. https://www.micromek.net/#prettyPhoto.

18. According to ISG (2017), the government of Rwanda is funding part of the operational costs.

19. "Greater Health Care Access Is One Drone Flight Away," https://www.villagereach.org/work/drones-for-health/.

20. For more details on the cost modeling tool used in this analysis, see MIT-Zaragoza, Transaid, and VillageReach (2011).

21. The average speed for a motorcycle was 27 km/h; average flight speed for a UAV was 40.5 km/h, including takeoff and landing, in addition to 15 minutes for the worker to load the UAV, change the battery, and launch the UAV at each location.

22. For example, the assumed volume capacity ranges from 0.4 liter for the low-price vendor to 230 liters for the high-price vendor.

23. For example, drone ports, landing pads or runways, land lease or land fees, and unmanned traffic management and navigation software.

24. Current methods use public transport, considered an "all-inclusive" cost. UAV costs are inclusive of capital investment and depreciation to enable comparison.

25. Current costs reflect operating costs only; no capital costs or depreciation are included. UAV costs do not include initial capital investment or depreciation to enable comparison of equivalents. Current methods use cold boxes and vaccine carriers. UAV uses packaged unit weights and dimensions and includes a buffer for appropriate packaging during the flight.

26. Current methods use public transport, considered an "all-inclusive" cost. UAV costs are inclusive of capital investment and depreciation to enable comparison of equivalents.

27. Current methods use public transport, considered an "all-inclusive" cost. UAV costs are inclusive of capital investment and depreciation to enable comparison of equivalents.

28. However, fixed-wing UAVs generally require runways or landing strips, which is not feasible at some of the hard-to-reach facilities. Moreover, although the small fixed-wing model has a long range, its weight cargo capacity is only 1.5 kg.

REFERENCES

Adeshokan, O. 2019. "In Africa, Drone Technology a Gateway to STEM. *Devex*, June 27. https://www.devex.com/news/in-africa-drone-technology-a-gateway-to-stem-95137#.XcbXcsvDHoY.gmail.

Amukele, T. K., J. Hernandez, C. L. H. Snozek, R. G. Wyatt, M. Douglas, R. Amini, and J. Street. 2017. "Drone Transport of Chemistry and Hematology Samples over Long Distances." *American Journal of Clinical Pathology* 148 (5): 427–35.

Amukele, T. K., J. Street, K. Carroll, H. Miller, and S. X. Zhang. 2016. "Drone Transport of Microbes in Blood and Sputum Laboratory Specimens." *Journal of Clinical Microbiology* 54 (10): 2622–25.

Aryal, M., and S. Dubin. 2019. "Opinion: Flying the Last Mile—Integrating Cargo Drones into Health Supply Chains." *Devex*, November 6. https://www.devex.com/news/sponsored/opinion-flying-the-last-mile-integrating-cargo-drones-into-health-supply-chains-94545#.XcbXOfq5FG8.gmail.

Department of Civil Aviation, MACRA, VillageReach, GIZ, and UNICEF. 2019. "Malawi Remotely Piloted Aircraft (RPA) Toolkit: A Guideline for Drone Service Providers and Implementers in the Development, Humanitarian and Research Fields." December. https://www.villagereach.org/wp-content/uploads/2019/12/Malawi-RPA-Toolkit-2019_December.pdf.

Dirks, W. 2017. "Evaluation of the Business Cases for Cargo Drones in Humanitarian Action: Qualitative Analysis." Final report. Medecins Sans Frontieres, Tokyo, Japan.

Drammeh, B., A. De, N. Bock, S. Pathak, A. Juma, R. Kutaga, M. Mahmoud, et al. 2018. "Estimating Tanzania's National Met and Unmet Blood Demand from a Survey of a Representative Sample of Hospitals." *Transfusion Medicine Reviews* 32 (1): 36–42.

Drones for Development. 2016. "Dr.One Proof of Concept. Executive Summary." November. IDI Snowmobile, Amsterdam.

Gavi. 2014. Soutien en espèces au renforcement du système de santé (RSS). June 3.

Global Laboratory Initiative. 2014. "Mycobacteriology Laboratory Manual." https://www.who.int/tb/laboratory/mycobacteriology-laboratory-manual.pdf.

Haidari, L. A., S. T. Brown, M. Ferguson, E. Bancroft, M. Spiker, A. Wilcox, R. Ambikapathi, V. Sampath, D. L. Connor, and B. Y. Lee. 2016. "The Economic and Operational Value of Using Drones to Transport Vaccines." *Vaccine* 34 (34):4062–67.

Hassanalian, M., and A. Abdelkefi. 2017. "Classifications, Applications, and Design Challenges of Drones: A Review." *Progress in Aerospace Sciences* 91: 99–131.

Hii, M. S. Y., P. Courtney, and P. G. Royall. 2019. "An Evaluation of the Delivery of Medicines Using Drones." *Drones* 3: 52. doi:10.3390/drones3030052.

ISG (Interagency Supply Chain Group). 2017. "Addressing Adoption and Sustainability of Unmanned Aerial Systems (UAS) in Public Health." Interagency Supply Chain Group.

ITF (International Transport Forum). 2018. "(Un)certain Skies? Drones in the World of Tomorrow." International Transport Forum, OECD, Paris.

JSI (John Snow, Inc.). 2018a. "Deliver the Future Phase II Preparation." Technical Report on Data Research, John Snow, Inc.

JSI (John Snow, Inc.). 2018b. "Implementation Plan for Transport Network Including UAVs." John Snow, Inc.

JSI (John Snow, Inc.). 2018c. "Network Assessment & System Design for Transportation of EID Samples and Test Results in Malawi." John Snow, Inc., Arlington, VA.

Juskauskas, T. 2019. "Drone Ecosystem in Malawi: Humanitarian Drone Corridor, UTM, African Drone and Data Academy." UNICEF Malawi, Presentation at Lake Victoria Challenge, May 7.

Knoblauch, A. M., S. de la Rosa, J. Sherman, C. Blauvelt, C. Matemba, L. Maxim, O. D. Defawe, et al. 2019. "Bi-Directional Drones to Strengthen Healthcare Provision: Experiences and Lessons from Madagascar, Malawi and Senegal." *BMJ Global Health* 4:e001541. doi:10.1136/bmjgh-2019-001541.

Machagge, M. 2018. "Medical Commodities Delivered by AAVs: Challenges & Opportunities." October 29.

Maisonet-Guzman, O. 2014. "Drones—The Next Development Game-Changer?" Devex, January 17. https://www.devex.com/news/drones-the-next-development-game-changer-82672.

Malawi Ministry of Health. 2016. "Integrated HIV Programme Report Quarter 3 (2016)." Ministry of Health, Lilongwe.

Malawi National Statistical Office. 2017. "Malawi Demographic and Health Survey 2015–16." Zomba, February.

Matemba, C. 2019. "Unmanned Aerial Vehicle (UAV) Cost Modeling for Commodity Delivery in Malawi." VillageReach, 2019 Global Health Supply Chain Summit, Johannesburg, November 21.

McCall, B. 2019. "Sub-Saharan Africa Leads the Way in Medical Drones." *The Lancet* 393 (10166): 17–18 .

McVeigh, K. 2018. "'Uber for Blood': How Rwandan Delivery Robots Are Saving Lives." *The Guardian*, January 2, 2018. https://www.theguardian.com/global-development/2018/jan/02/rwanda-scheme-saving-blood-drone.

Mikou, M., J. Rozenberg, E. Koks, C. Fox, and T. Peralta Quiros. 2019. "Assessing Rural Accessibility and Rural Roads Investment Needs Using Open Source Data." Policy Research Working Paper 8746, World Bank, Washington, DC.

MIT-Zaragoza, Transaid, and VillageReach. 2011. "Framework on Distribution Outsourcing in Government-run Distribution Systems." http://www.villagereach.org/wp-content/uploads/2011/05/Framework-on-Outsourcing-Reference-Material-for-Nigeria.pdf.

Mulamula, G. 2018. "Building Wakanda: The Connected Village, the Smart City, and IoT Technology – Connecting Villages and Islands in Lake Victoria for Better Health." LVC Symposium, July 18.

Müller, S., C. Rudolph, and C. Janke. 2019. "Drones for Last Mile Logistics: Baloney or Part of the Solution?" *Transportation Research Procedia* 41: 73–87.

Oxford Infrastructure Analytics Ltd. 2018. "Transport Risk Analysis for the United Republic of Tanzania: Systemic Vulnerability Assessment of Multi-Modal Transport Networks." Final Report. Oxford Infrastructure Analytics Ltd, Oxford, U.K.

Pai, M., S. G. Schumacher, and S. Abimbola. 2018. "Surrogate Endpoints in Global Health Research: Still Searching for Killer Apps and Silver Bullets?" *BMJ Global Health* 3 (2): e000755.

Phillips, N., C. Blauvelt, M. Ziba, J. Sherman, E. Saka, E. Bancroft, and A. Wilcox. 2016. "Costs Associated with the Use of Unmanned Aerial Vehicles for Transportation of Laboratory Samples in Malawi." VillageReach, Seattle.

Rabien, D. 2018. "Logistics of Drone Delivery: Deliver Future (Drone X)." Lake Victoria Challenge, GIZ, Bonn.

Rosen, J. W. 2017. "Zipline's Ambitious Medical Drone Delivery in Africa." *MIT Technology Review*, June 8, 2017. https://www.technologyreview.com/s/608034/blood-from-the-sky-ziplines-ambitious-medical-dronedelivery-in-africa/.

Rugambwa Bwanakunu, L. 2018. Presentation made at the Lake Victoria Drone Conference, Mwanza, Tanzania. Medical Stores Department (MSD), October.

Rupani, S. 2017. "UAVs in Supply Chains: Thinking about UAVs as a Transport Mode." LLamasoft, Inc., Global Impact Team.

Small, P. 2017. "Drones in Madagascar." Global Health Institute, June 30.

TACAIDS (Tanzania Commission for AIDS), ZAC (Zanzibar AIDS Commission), NBS (National Bureau of Statistics), OCGS (Office of the Chief Government Statistician), and ICF International. 2013. *Tanzania HIV/AIDS and Malaria Indicator Survey 2011-12*. Dar es Salaam, Tanzania: TACAIDS, ZAC, NBS, OCGS, and ICF International.

Tanzania NBS (National Bureau of Statistics). 2013. "2012 Population and Housing Census." National Bureau of Statistics, Dar es Salaam.

Thiels, C., J. M. Aho, S. P. Zietlow, and D. H. Jenkins. 2015. "Use of Unmanned Aerial Vehicles for Medical Product Transport." *Journal of Air Medical Transport* 34 (2): 104–8.

Ukerewe District Council. 2017. "2017/18 Comprehensive Council Health Plan."

UNICEF. 2016. "Malawi Tests First Unmanned Aerial Vehicle Flights for HIV Early Infant Diagnosis." UNICEF, Lilongwe, Malawi. https://www.unicef.org/media/media_90462.html.

UNICEF. 2018. "UNICEF Malawi Unmanned Aerial Systems." Lake Victoria Challenge, Tanzania. October 29.

UNICEF Malawi. 2019. "Call for Expressions of Interest (EOI): Research, Evaluation, and Market Analysis Services Related to Multi-Purpose Unmanned Aerial Systems (UAS)."

UNICEF Tanzania. 2018. *Annual Report 2018*. Dar es Salaam, Tanzania: UNICEF Tanzania.

UNICEF and WHO (World Health Organization). 2014. Évaluation de la Gestion Efficace des Vaccins en République Démocratique du Congo. September 22 – October 28.

UPDWG (UAV for Payload Delivery Working Group). 2019. "Integrating Drones into Immunization Supply Chains." UAV for Payload Delivery Working Group. November 7.

USAID (United States Agency for International Development). 2013. "Last Mile Costs of Public Health Supply Chains in Developing Countries. Recommendations for Inclusion in the United Nations OneHealth Model." Deliver Project, Task order 4, John Snow, Inc., for United States Agency for International Development.

USAID (United States Agency for International Development). 2017. *Unmanned Aerial Vehicles Landscape Analysis: Applications in the Development Context*. Global Health Supply Chain Program-Procurement and Supply Management. Washington, DC: Chemonics International Inc. for United States Agency for International Development.

USAID (United States Agency for International Development). 2018. *UAVs in Global Health: Defining a Collective Path Forward*. Center for Accelerating Innovation and Impact (CII). Washington, DC: United States Agency for International Development.

VillageReach. 2019a. "Drones for Health: Boosting Access to Lifesaving Products." VillageReach.

VillageReach. 2019b. "Toolkit for Generating Evidence around the Use of Unmanned Aircraft Systems (UAS) for Medical Commodity Delivery." Version 2, November. VillageReach.

Wakefield, J. 2019. "The Airport That Welcomes Drone Flights." BBC News. January 26.

WEF (World Economic Forum). 2018. *Advanced Drone Operations Toolkit: Accelerating the Drone Revolution*. Geneva: World Economic Forum.

WHO (World Health Organization). 2005. "Manual on the Management, Maintenance and Use of Blood Cold Chain Equipment: Safe Blood and Blood Products." World Health Organization, Geneva.

WHO (World Health Organization). 2015. "How to Use Passive Containers and Coolant-Packs for Vaccine Transport and Outreach Operations." In WHO Vaccine Management Handbook, WHO/IVB/15.03(Module VMH-E7-02.1), WHO, Geneva.

WHO (World Health Organization). 2018. *Global Tuberculosis Report 2018*. Geneva: World Health Organization.

Wright, C., S. Rupani, K. Nichols, Y. Chandani, and M. Machagge. 2018. "What Should You Deliver by Unmanned Aerial Systems? The Role of Geography, Product, and UAS Type in Prioritizing Deliveries by UAS." JSI Research & Training Institute, Inc.

Zipline. 2018. "How Zipline Works." http://www.flyzipline.com/service/.

2 Food Aid Delivery

THE BIG PICTURE: THE REGIONAL DEMAND FOR FOOD AID

Food security remains a major challenge in East Africa, with adverse climate conditions exacerbating the situation, leaving millions in need of urgent assistance. The number of people that are undernourished, as defined by the Food and Agriculture Organization, is on the rise in Kenya, Rwanda, Tanzania, Uganda, and Zambia; as a share of the population, the prevalence of undernourishment has increased in Kenya, Rwanda, and Uganda. Significant areas of Ethiopia, Somalia, South Sudan, and Sudan are characterized as extremely food insecure (map 2.1). According to UNICEF, in Ethiopia alone, nearly 8 million people require food assistance, and 370,000 children require treatment for severe acute malnutrition. Seasonal climate-related floods and droughts affect specific subregions, compounding acute food insecurity, malnutrition, and water shortages, mostly in pastoral and highland areas. Undernutrition is the cause of about 28 percent of child deaths in the country. In Tanzania, more than 600,000 children younger than age five are estimated to suffer from acute malnutrition (UNICEF 2017).

In mid-2019, more than 1.8 million people in South Sudan needed nutrition-related aid according to the United Nations Office for the Coordination of Humanitarian Affairs. Further deterioration of food security was expected during the postharvest period, with 5.5 million people projected to be facing crisis-level acute food insecurity or worse (UNICEF 2019). In November 2019, 490,000 children in South Sudan were affected by floods, the consequences of which were felt long after the water subsided. Damaged crops and grazing land submerged in water hamper access to food for many children and their families, making assistance from organizations such as the World Food Programme (WFP) and UNICEF essential.

In Rwanda, which faces transportation and logistics issues common to land-locked countries, more than 38 percent of the population continues to live below the poverty line and almost one-fifth is food insecure. Levels of stunting among young children remain at 35 percent according to the WFP. As reported by UNICEF, the likelihood of stunting among 18–23-month-old children is a staggering 49 percent, with children more likely to be stunted if they live in very

MAP 2.1

Food insecurity outlook, February–May 2016

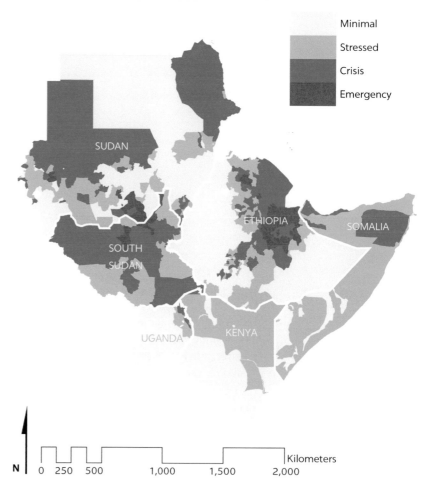

Source: Data from 2016 East Africa Famine Early Warning Systems Network Food Security Outlook.

poor households or in rural areas. The high prevalence of stunting is a result of several factors, including inadequate diet. Shocks induced by climate change, such as the recent successive droughts, have become a major driver of food insecurity (WFP 2018).

In 2016–18, nearly 21 million metric tons (mt) of maize, in addition to 8.4 million mt of millet/sorghum and 10.9 million mt of wheat, were consumed annually by Burundi, Djibouti, Ethiopia, Kenya, Rwanda, Somalia, South Sudan, Tanzania, and Uganda,[1] with consumption exceeding production by a large margin in most countries and the gap being filled by a combination of imports and humanitarian aid deliveries. Poor infrastructure and high trade costs are often identified as critical constraints by local producers, especially in landlocked countries (Burundi, Rwanda, Uganda). More than 110 million people, or about 75 percent of the total population in the East African Community[2] countries alone, do not have access to the all-season road network within 2 kilometers (km) of their home (World Bank 2018). Climate change will further reduce connectivity, raising the potential for food shortages, spikes in food prices, and economic shocks to vulnerable areas (Cervigni et al. 2017). For example, the food-specific consumer price index in South Sudan reached 2,584 in January 2018 compared with the base level of 100 five years earlier, according to FAOSTAT, as a result of various shocks that affected food availability.

The WFP is the largest supplier of humanitarian food aid to East Africa, including from local and external production centers. In 2016, the WFP delivered 1.13 million mt to Burundi, Djibouti, Ethiopia, Kenya, Rwanda, Somalia, South Sudan, Tanzania, and Uganda combined, and in 2017 the delivered volume of food increased to 1.20 million mt. Between 2016 and 2017, South Sudan's and Uganda's shares of all food delivered increased (figure 2.1).

To predict how overall demand for food (including both nonhumanitarian and humanitarian) will evolve in the region over the next two decades, the IMF *World Economic Outlook* 2018 country population data and US Department of Agriculture/IndexMundi data on historic consumption of maize, millet/sorghum, and wheat in each of the countries can be used as inputs to estimate the past relationship between growth in population and growth in demand for specific food crops by applying the so-called population multiplier. The multiplier is calculated by dividing the compound annual growth rate (CAGR) of the consumption of food staples by the CAGR of population over that same period. The population CAGR derived for 2012–17 ranges from 1.66 percent in Djibouti to 3.38 percent in Uganda; it was 2.50 percent in Rwanda and 3.06 percent in South Sudan. The resulting base scenario prediction shows that by 2038 these countries' total demand for maize will be about 27 percent higher than in 2018, and demand for millet/sorghum will increase by 30 percent compared with 2018, with the largest percentage growth in demand predicted in South Sudan (51 percent). The demand for wheat is predicted to increase by 157 percent by 2038.

FIGURE 2.1

World Food Programme food deliveries to the East Africa region, by recipient country, 2016–17

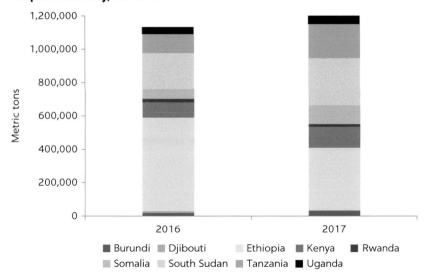

Source: Data obtained from World Food Programme.

Aid organizations such as the Food and Agriculture Organization, UNICEF, and the WFP will likely need to supply a significant share of this future demand for food, especially in the region's poorest countries, such as South Sudan. For example, WFP shipments in 2016–17 accounted for 3 percent of South Sudan's overall reported maize consumption and as much as 42 percent (2016) and 60 percent (2017) of its millet/sorghum consumption. Identifying potential opportunities for transport cost savings in responding to the region's enormous food aid needs is essential, given the limited funds available. For example, in Ethiopia, UNICEF in 2018 had only $49.1 million available against the $123.8 million appeal (UNICEF Ethiopia 2019).

The WFP is also a major supporter of regular school feeding programs across the developing world. In 2018, it implemented or supported school feeding programs in 64,000 schools across 71 countries, providing school meals or snacks to 16.4 million children, of whom 2.44 million were living in East Africa. The WFP helps provide daily school meals in 104 schools in Rwanda, in four vulnerable and food insecure districts in the country's Southern and Western provinces. The Home-Grown School Feeding (HGSF) program, funded primarily by the United States Department of Agriculture, is implemented by the WFP on behalf of the Rwandan government and is one of the chief WFP activities in the country: it is one of the main pillars of the five-year strategic plan (Country Strategic Plan 2019–23) launched by the

MAP 2.2

Population density in Rwanda

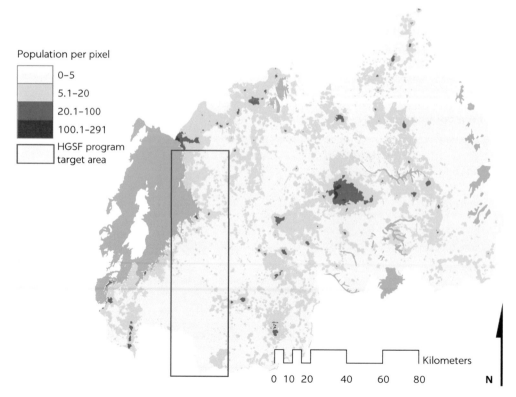

Source: WorldPop population data, 2020.
Note: HGSF = Home-Grown School Feeding.

MAP 2.3

Poverty rates in Rwanda

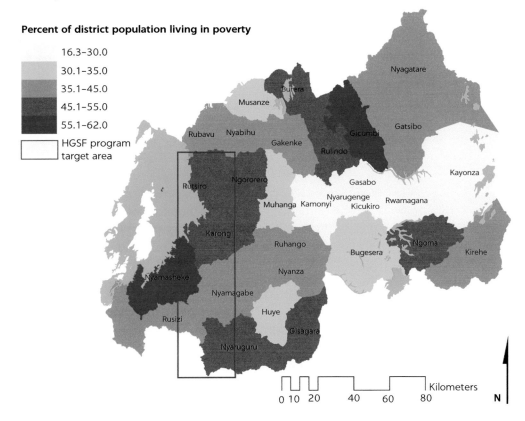

Percent of district population living in poverty

- 16.3–30.0
- 30.1–35.0
- 35.1–45.0
- 45.1–55.0
- 55.1–62.0
- HGSF program target area

Source: NISR 2015.
Note: HGSF = Home-Grown School Feeding.

WFP Rwanda country office. The program specifically targets 40,000 children in the Nyamagabe and Nyaruguru Districts in the Southern Province and 43,000 children in the Karongi and Rutsiro Districts in the Western Province , corresponding to some of the lowest population densities and highest rates of poverty in the country according to the 2013/14 Rwanda Poverty Profile Report, exceeding 45 percent in Nyaruguru, Karongi, and Rutsiro (NISR 2015) (maps 2.2 and 2.3). A total of 6,720 kilograms (kg) of food each day is consumed by the 40,000 children in the program in the Nyamagabe and Nyaruguru Districts; the corresponding daily figure for the Karongi and Rutsiro Districts combined is 5,805 kg (WFP Rwanda 2019). In fiscal year 2017, the program delivered 60 net mt of vegetable oil alone (to the Southern Province districts) and 460 net mt of Supercereal (to the Western Province districts).

PRESENT COSTS AND MODALITIES

Food aid to South Sudan

Current food aid demand data suggest that transport costs in South Sudan are significantly higher than in the rest of East Africa, driven mainly by lack of a year-round, all-weather road network. Overall shipment costs handled by the

WFP just for delivering maize to the final destinations in South Sudan in 2017 amounted to $10.67 million, the highest transport cost among the East Africa region's recipient countries. The WFP supply chain for emergency food aid in South Sudan, unlike in the rest of East Africa, is based to a large extent on aviation—aircraft leased from external operators and operated through the WFP's Common Services Platform for Aviation & Logistics. Aviation, that is, airplanes and helicopters, is the most expensive mode for freight delivery per ton in the country and is used for delivering food aid from larger population centers such as Juba to airdrop zones near towns and villages (figure 2.2). Many of the smaller villages are inaccessible during some parts of the year because of climate-related events such as floods, resulting in fewer people that can be served.

Aggregated data on food airdrop operations performed by the WFP in South Sudan in 2018 and 2019 were provided for the purposes of this study by field-based WFP staff. In 2018, 2,299 airdrops using aircraft were made in South Sudan from loading locations in Juba (South Sudan) and Gambela (Ethiopia), for a total volume of 60,667 mt. The food aid from the Juba and Gambela loading stations is airdropped to more than 90 airdrop zones across South Sudan, located near inhabited villages, with individual villages being served by more than one airdrop zone (map 2.4). In 2019, the overall volume of food airdropped decreased significantly compared with 2018, to about 20,500 mt, as a result of supply chain optimization (aviation is the most expensive delivery option) and changes in the situation on the ground having to do with food availability and demand. The average payload per airdrop in 2019 was 30.1 mt, requiring the use of relatively large cargo aircraft.

The average cost per mt of airdrop, excluding the cost of the food itself but including all cost components of delivery, was $1,830 in 2018–19. The transport operating cost for the main types of food delivered ranged from $432 per flight hour (FH) per mt to $590 per FH/mt for cereals and pulses and from $682 per FH/mt to $932 per FH/mt for cereals and oils. The cost range for each food category is due to variations in local fuel prices, the cost of insurance for current conditions, and other factors.

FIGURE 2.2

Schematic of WFP food aid delivery network in South Sudan

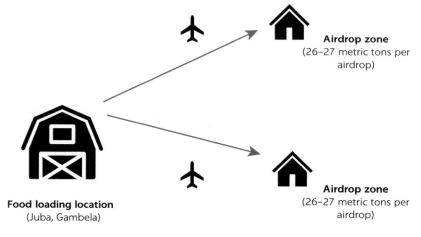

Airdrop zone
(26–27 metric tons per airdrop)

Airdrop zone
(26–27 metric tons per airdrop)

Food loading location
(Juba, Gambela)

Source: Interviews with World Food Programme (WFP) staff.

MAP 2.4

Food aid loading and airdrop zones in South Sudan

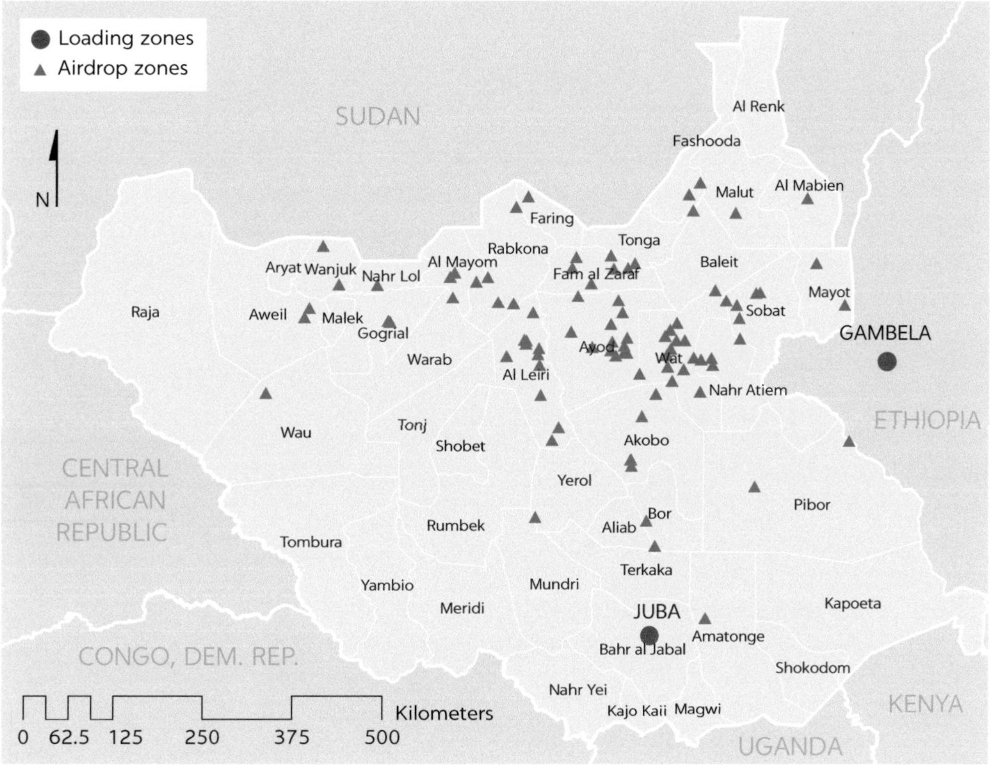

Source: World Food Programme South Sudan data.

School feeding program in Rwanda

Lack of all-weather road connectivity makes food delivery for Rwanda's school program challenging, particularly when it comes to the "last mile." The food commodities distributed to schools under the HGSF program are sourced both as in-kind food commodity transfers from the United States and as home-grown local and regional purchases. The Western Province districts targeted by the HGSF receive in-kind Supercereal and sugar, and the Southern Province districts receive home-grown, locally purchased maize meal, sugar, vegetable oil, and salt.

Deliveries to schools are made every three months (once per term) to minimize delays during the rainy season; still, the program report for the third quarter of fiscal year 2018 notes delivery delays for both Supercereal and vegetable oil owing to challenges such as limited access to schools during the rainy season and poor rural road infrastructure. Food quality is also being managed by transferring commodities among schools to ensure that food with best-before use dates is evenly distributed, thus preventing waste due to expired commodities (WFP Rwanda 2019). The most challenging area for food deliveries, with the highest inland transport costs per mt delivered, are the districts near Lake Kivu, where road quality is poor; major roads are absent or are not paved (map 2.5), and it is frequently difficult for the organization to identify transporters willing to accept delivery contracts. Last-mile deliveries are particularly difficult, with food aid sometimes carried on foot by the transporters because of impassable roads.

MAP 2.5

Road network in Rwanda

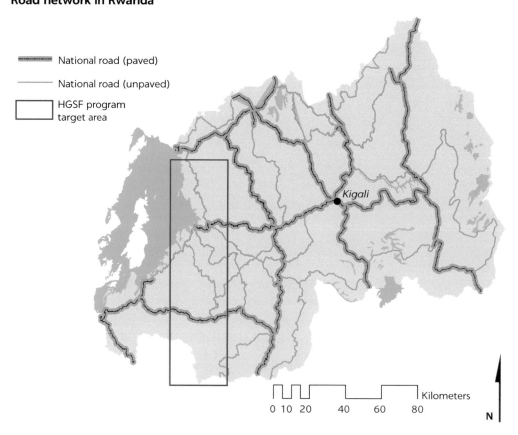

Source: Rwanda Transport Development Agency (City of Kigali, 2018).
Note: HGSF = Home-Grown School Feeding.

The all-in inland transport costs from Kigali to three of the districts under the HGSF program, as reported in 2018–19, are high compared with the transport costs incurred by the WFP for food deliveries to other provinces in Rwanda, reaching about $28 per mt for Nyamagabe and about $19.5 per mt in Karongi and Rutsiro. Delivery costs to Nyaruguru are lower, at about $14 per mt. Taking into account the road distances traveled from Kigali, the cost per ton-km is approximately $0.18 to deliver food to schools in Nyamagabe, about $0.155 to Rutsiro and Karongi, and $0.08 to Nyaruguru.

INTEGRATING DRONES INTO EXISTING FOOD AID DELIVERY SYSTEMS

Supply chains and route types in which unmanned aerial vehicles (UAVs) can add value

The food aid delivery cases in Rwanda and South Sudan confirm that drones can potentially augment, rather than replace, the existing aviation- or land transport–based supply chains. Any use of drones to transport food aid would have to consider best practices related to food airdrops as developed by the WFP and other humanitarian organizations. Best practices include having staff on the ground in the drop zone. Airdrops without a presence on the ground are a measure of last resort because of, among other reasons, the immense difficulties of ensuring that the food is distributed to those most in need. Drone deliveries would not change this vital human element of the food distribution pipeline (Soesilo et al. 2016).

In addition to potential cost advantages in certain contexts, the rationale for integrating drones into food aid delivery supply chains may lie in their flexibility and speed, especially compared with road transport. The WFP's business model in many countries is primarily based on real-time, quick shipping (anticipated no more than two to three weeks in advance). Therefore, reliability of delivery times is more important than average delivery lead times, and unexpected delays have significant financial cost implications. Analysis of the data for the trip legs handled by the WFP for the maize purchased in 2017 for East African countries indicates that deviations between planned lead time and actual lead time tend to be very large, even for within-country trip legs that do not involve a border crossing.

In South Sudan, as in most recent food aid recipient countries, cargo drones will likely only step in when and where regular aviation modes fail or underperform. Each year, more than 12,300 people in South Sudan are exposed to floods, and humanitarian organizations have difficulty delivering aid to them (INFORM 2015) (map 2.6). As reported by UNICEF, the ongoing crisis in South Sudan has forced large numbers of women and children to move to

MAP 2.6

Physical exposure to floods in South Sudan

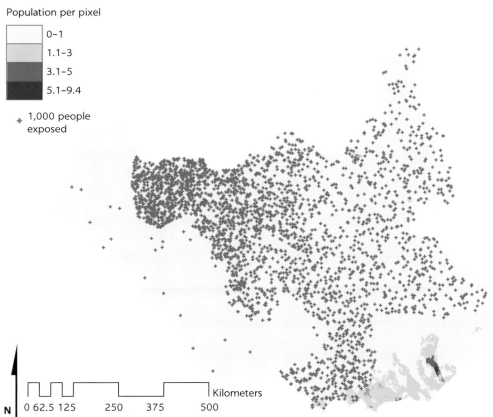

Population per pixel

	0–1
	1.1–3
	3.1–5
	5.1–9.4

+ 1,000 people exposed

Sources: WorldPop population data, 2020; INFORM (2015) hazard risk data obtained from Humanitarian Data Exchange.

inaccessible areas where basic lifesaving services are unavailable. Manned aircraft are not always able to reach these locations to deliver much-needed humanitarian aid, but drones may provide a feasible alternative. However, any use of cargo drones would have to ensure high accuracy of location delivery if parachutes are used and very high accuracy of flight paths. The current reality, too, suggests that some hurdles remain to be resolved for drone-based food deliveries to commence. Several platforms have been established but none of the operators is yet formally certified.

Large payloads are the norm in food aid delivery unless there are exceptional emergency circumstances. Generally, drones with payload capacity of at least 80–100 kg, similar to the payload of a Yamaha helicopter and capable of serving the food needs of at least a small village, are considered to be the minimum for being useful in the food aid deliveries performed in East Africa by the WFP. However, smaller payload drones could be used in true emergencies, as well as in UNICEF's operations, in which smaller volumes are sometimes airlifted. An average regular aircraft currently used by the WFP to deliver food aid in South Sudan has the capacity to airlift 40 tons of cargo. It would take many medium-size cargo drones to carry the same payload. It is therefore not feasible for drones to replace this overall airlift capacity, even just considering the requirements that would be put on local airport capacity to release so many vehicles. It is more likely that only up to 20 percent or so of the overall food aid delivered by the WFP in South Sudan could be served by drones. As noted, the space for drone deliveries would be in true emergency situations and for deliveries to smaller villages.

Although multiple small-payload drones could theoretically replace one small cargo aircraft, the greater opportunity for food aid delivery would come from enabling heavy cargo drones. A drone would need a payload capacity of at least 100 kg (enough to serve a small village) to add value in WFP operations in the South Sudan context; a payload capacity of 1 ton would be required to deliver food to a large village. The current range and payload capacity of most cargo drones limit where drones could add value to the servicing of insecure areas that are within 100 km; however, adopting the technology could also have broader positive impacts on overall logistics efforts (Soesilo et al. 2016). Cargo drones, especially those the size of regional or intercontinental cargo aircraft, may also lower the cost per ton through other factors such as lower personnel cost from a switch to ground-based crews, increased flexibility of flight schedules, and lower fuel costs (Hoeben 2014).

Some drone manufacturers are already moving in the heavy cargo direction with humanitarian aid emerging as the front-runner of potential use cases. The minimum payload requirements expected by humanitarian aid organizations such as the WFP and UNICEF are already met by a range of cargo drones currently being developed, such as the Flyox and ZHZ/United Aircraft vehicles. The cargo drones currently in development range from those with a 300 kg payload and 500 km range to those with a 15,000 kg payload and a range of up to 15,000 km. A number of global companies are experimenting with drone delivery tailored to special use cases, including the development of drones for large payloads (about two tons). Small freight delivery drones for payloads of up to 5 kg are being piloted in various countries in everyday conditions, such as by Amazon and DHL; however, humanitarian services are currently the most prominent use cases for freight drones in the small to mid-size payload category (ITF 2018).

The drone manufacturers are counting on reaching comparable capacity to small cargo planes while offering unique advantages that bring efficiency and around-the-clock operation potential. With regard to payload, the Flyox I drone appears to be comparable to the Cessna Caravan and is nearly competitive with other types of small aircraft such as the Dornier 288 and the Fairchild Metroliner (Gadhia 2017). The Flyox drone, intended to be used by Astral Aerial Solutions, the Kenya-based cargo drone company, is operation-ready in an hour, and, unlike small aircraft, is capable of autonomous takeoff and landing both during day- and nighttime. The large cargo drones also have fewer (albeit not zero) crew requirements. The Flyox cargo drone can theoretically carry a cargo payload of 1.8 mt over a range of 1,200 km. Astral Aerial Solutions is also exploring the use of the medium-size Falcon F250 drone, able to carry up to 250 kg of cargo more than 150 km. Similarly, Boeing has recently introduced a cargo drone that can carry up to 227 kg.

Nevertheless, data from Rwanda confirm the potential of small cargo drones to perform the last-mile delivery. In Rwanda, drones have the potential to provide value as a component of the existing land-based transport delivery system; they can provide last-mile access to remote schools where even small trucks have difficulty traveling, especially in the rainy season, because of the poor quality of the smaller access roads despite the various ongoing road con- struction and rehabilitation programs (such as by the World Bank and the Japan International Cooperation Agency). No actual drone-based operations have yet commenced in the WFP's program in Rwanda, including because of the high minimum payload requirements for drones to be used in the delivery of food aid.

Potential cost and practical implications

Food aid to South Sudan

Heavy cargo drones are currently at relative cost parity with manned aircraft, although costs are decreasing and capabilities are increasing rapidly. Cargo drones of the one or more ton payload capacity required for most WFP needs in South Sudan currently cost between $1 million and $2 million, and the opera- tional cost is projected to be very close to that of regular aircraft operations. However, new drone manufacturers are entering the market that may be able to offer a significantly lower up-front price. For example, the JDY-800 drone developed by the China-based company JD and unveiled in 2018 will be able to fly at speeds of more than 200 km/h and carry up to 840 kg of cargo; the drone is expected to cut logistics costs by 70 percent compared with ground-based shipment modes (China Daily 2018).

The current transport cost incurred for delivering food aid using manned air- craft, including the empty backhaul, is estimated to be about $4.58 per mt-km. The analysis of the costs that may be associated with using UAVs to meet a share of the food aid delivery needs served by the WFP in South Sudan considers data on the food volumes delivered in 2019, given the changed reality on the ground from what prevailed in 2018.[3] The analysis assumes that those villages closer to Juba were airdropped from Juba and those closer to Gambela were airdropped from Gambela. It is also assumed that each of the approximately 90 airdrop zones receive equal volumes,[4] equivalent to about 220 mt per year. As noted, the transport costs currently incurred by the WFP for the main types of food delivered in South Sudan using conventional aircraft average about $660 per FH per mt, and the overall average cost per airdrop per mt is $1,830,[5] which covers the round trip (backhaul empty) of an average distance from the closest loading zone to the airdrop destination of approximately 400 km each way. WFP Aviation, the cargo arm of the WFP's transport and logistics operations, currently leases aircraft and operates on the model of chartered flights to deliver food. Different contracted operators are typically able to offer different prices per mt airdropped.

Some studies show that under certain combinations of demand and time frames, cargo drones can be more cost-effective than manned aircraft. Insights gathered for this study from both humanitarian aid organizations and drone developers suggest that the real competitors to cargo drones are manned aircraft. The rule-of-thumb cost for air transport is approximately $0.60 to $0.90 per ton-km, or at least double the per-ton-km cost of trucking and orders of magnitude larger than the cost of rail or waterway transport (Rupani 2017). Van Groningen (2017) analyzes the cost efficiency of a long-haul drone transporting a 5,000 kg payload weekly between Stuttgart, Germany, and Shenzhen, China, and compares it with other possible modes of transport. The study concludes that, for this specific scenario, and when shipping automotive goods, a cargo drone would be 17 percent more cost-effective than a Boeing 777. However, the study also shows that the achieved efficiency largely depends on the amount of freight demand and its timing, with a minimum of 1,250 kg of high-value payload transported between Stuttgart and Shenzhen when competing with a Boeing 777 or 500 kg when competing with rail.

Indeed, when using large cargo aircraft as the counterfactual, drones available in the market can be cost competitive on a per mt-km basis. Although mindful of the rapid advances in the cargo drone market and the potential for significantly lower capital and operating costs that can, or could soon, be offered by individual drone manufacturers and operators, the current analysis is based on the technical and cost parameters of the Flyox I drone, which can reliably carry 1 mt of humanitarian aid food.[6] Remotely piloted from a ground station, the drone can take off and land not only on unpaved runways but also on water. Its operation and maintenance cost[7] is an estimated $220 per FH,[8] although it is likely to decline further and depend to some extent on ground operating costs, which may differ across countries. Assuming a cargo payload of 1 mt and a cruising speed of 200 km/h, the cost is thus about $1.10 per mt-km one way or $2.20 per mt-km if the fact that the drone must travel back empty the same distance is taken into account (figure 2.3). According to these estimates with large cargo aircraft as the counterfactual, drones able to carry 1 mt at $220 per FH are thus cost competitive on a per mt-km basis in the specific context of the WFP's operations in South Sudan.

Despite the estimated lower costs per mt-km, the use of drones would likely not be justified for all food aid deliveries in South Sudan. One major reason is the difference in payload capacity between the current aircraft used by the WFP (more than 30 mt) and cargo drones with payload capacity of 1 mt. These payload differences, in turn, imply large differences in the number of airdrops that would have to be released from airfields; for example, to be able to deliver 3,000 mt of food, a plane carrying the average 2019 payload per airdrop of 30 mt would require 100 trips, whereas a drone with a payload of 1 mt would need 3,000. Based on the insights of WFP staff working in South Sudan, drones could most likely be used as an alternative transport solution for up to 20 percent of all airdrop deliveries in South Sudan, equivalent to about 4,100 mt annually if using

FIGURE 2.3

Estimated round-trip transport cost to airdrop food in South Sudan using manned aircraft vs. cargo drones

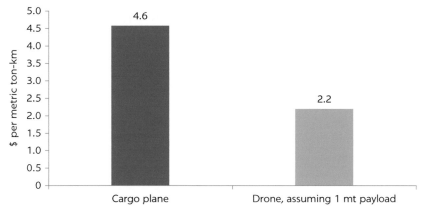

Source: World Bank calculations based on World Food Programme data.
Note: mt = metric ton.

2019 deliveries as a benchmark. Specifically, drone-based airdrops would focus on destinations or situations where they would provide a real practical advantage or be the only practically feasible solution, for example, because of climate or security conditions, when manned aircraft would not be able to physically reach the airdrop destinations and safely return. Given that almost all airdrop locations are about equally likely to be affected by flooding (comparing maps 2.4 and 2.6) and intensified conflict, it could be assumed that the average one-way distance for the 20 percent of food deliveries to be reserved for drones would be the overall average across all airdrop destinations, or about 400 km. The annual transport cost savings alone, over and above the benefits of food aid safely reaching the intended recipients, would then amount to nearly $3.9 million.

Abnormal emergencies in the food aid delivery supply chain covered by smaller aircraft are the best case for drones in this use case. Another scenario in which drones may provide practical advantages compared with manned airplanes could be in those cases, such as during true emergency conditions, when much smaller quantities of food would need to be delivered to individual destinations and therefore small manned aircraft (such as a Cessna Caravan or something similar) would be the real counterfactual to drones. In this case, the payloads of the aircraft and the drones would be much more comparable, as would be the number of airdrops required.

School feeding program in Rwanda

Based on current inland transport costs for ensuring food delivery to schools, drone-based transportation from Kigali at or below about $28 per mt delivered would generate cost savings in at least one of the target districts (Nyamagabe) if considering the all-in delivery cost from the origin to the destination. If considering transportation from Kigali as an uninterrupted straight-line flight, the all-in cargo drone transport cost that would be at parity with current transport modes would be approximately $0.31 per ton-km for deliveries to Nyamagabe, the currently most-expensive-to-service district. The all-in price parity for deliveries to the other districts would range from $0.14 to $0.25 per mt-km.

These figures are well below what is currently reasonable to expect in the large cargo (about 1 mt payload) drone market, where the cost per mt-km (allowing for backhaul, assuming it is empty) is greater than $2. Therefore, on a cost basis alone, and especially for deliveries all the way from Kigali, the use of cargo drones is likely not justifiable. However, given the current spatial distribution and condition of major roads, and based on insights of WFP staff working in Rwanda, it is likely that the current, road-based marginal cost per km varies widely along each of the routes from Kigali to the final destinations, and that cargo drones could be cost competitive for the segments that large trucks have the most difficulty traveling. As illustrated in map 2.5, although most of the initial route from Kigali to any of the four HGSF target districts is along paved major roads, the last segments, at least 10–20 km, depending on the destination, must be completed using smaller or unpaved roads. These last-mile segments are likely not only much more expensive for transport companies on a per-km basis but are entirely impossible to travel by motorized means during parts of the rainy season, putting the delivery of food products to the target schools at risk.

For the schools in Nyamagabe, for example, the current high average trucking cost of $28 per mt or $0.179 per mt-km conceivably varies along the route from Kigali, given that the last 20 km after turning off paved National Road 6 must be completed on an unpaved road that becomes difficult or impossible to traverse during certain months of the year. Thus, it could be assumed that the first 136 km from Kigali are associated with a much lower cost per mt-km than the overall route average—a more reasonable $0.12–$0.13 per mt-km—while the last 20 km to Nyamagabe incur a cost of $0.50 per mt-km (in addition to the risk of the cargo potentially not being delivered at all). In that case, although the first 136 km of the route entail a transport cost of about $17.50 per mt, the last 20 km cost $10 per mt. In comparison, using a Flyox-type cargo drone instead of a truck for the last 20 km road distance (about 13 km air distance) would incur a total cost of about $28 per mt. The total route cost from Kigali would thus increase from the current $28 per mt to about $46 per mt. The cost difference is quite substantial, and, based on up-front transport costs alone, cargo drones would currently not be competitive with trucks even for last-mile delivery. However, the additional up-front cost may be acceptable in some circumstances, given that the use of drones for last-mile delivery would reduce overall cost uncertainty, the risk of late delivery or nondelivery of food for thousands of school children, and the cost of holding food inventory at the school level.

In the analysis of the future cost-effectiveness and potential advantages of drones versus land-based transport, in Rwanda and elsewhere, it is important to consider that changes—albeit at different rates—are occurring not only in the cargo drone market but also in the current network and modes of ground transport. Despite some remaining uncertainty about implementation, several transport improvement projects in the region are underway that are likely to improve overland transport on at least some of the routes currently used for WFP shipments. For example, in Rwanda, the World Bank's Lake Victoria Transport Program Phase 1, approved in 2017, expects to reduce travel time on the Kigali-Gasoro road to 1 hour and 45 minutes by 2023, down from 2 hours and 45 minutes in 2017, and to reduce transport costs for road users by 10 percent (De Serio 2019). In addition, the Rwanda Feeder Roads Development Project, in implementation since 2014, is helping to rehabilitate 720 km of rural roads; by 2022, the project expects to connect 50 percent of the target region's rural population to an all-season road, up from 15 percent in 2012, and to cut travel time per km by two-thirds (Taban 2019).

THE HUMAN IMPACT

There is a clear humanitarian case to be made for integrating drones into food aid delivery systems to address the needs of the more vulnerable. The advantage of UAVs in South Sudan—and possibly in other parts of Sub-Saharan Africa, such as the Sahel or Lake Chad area, that are also characterized by fragility, conflict, and climate vulnerability—lies primarily in humanitarian organizations' increased ability to serve populations in the most vulnerable locations where *traditional types of aircraft are not able to fly safely. The human impact is poten-*tially enormous, measured in many additional lives saved and reduced long-term impacts of severe malnutrition. Improvements in the efficiency of the supply chain for food aid deliveries would help achieve the WFP's overall goal of "zero hunger," which corresponds to Goal 2 of the 17 Global Goals for Sustainable Development adopted in 2015 to improve people's lives by 2030.

Children who suffer from malnutrition can experience its effects for life, and returns on investments in nutrition and health have effects on society as a whole. In the short term, malnutrition results in higher rates of mortality and decreased cognitive, motor, and language development. Long-term consequences include an increase in noncommunicable diseases, low school performance, and decreased work capacity. Ultimately, these outcomes affect society as a whole, limiting the ability of younger generations to advance out of poverty. Having food at school every day can mean not only better nutrition and health, but also increased access to and achievement in education. It is also a strong incentive to consistently send children to school. The regular availability of food is especially important for allowing girls to remain in school. School feeding has been found to be an effective intervention for boosting student learning, especially in Burkina Faso, Kenya, and Senegal (Bashir et al. 2018). Studies have shown that these types of programs can increase enrollment by an average of 9 percent, and by 12 percent among girls (Snilstveit et al. 2015). The HGSF program supported by the WFP in the Southern and Western Provinces of Rwanda fulfills a significant portion of daily nutritional requirements of primary school children, reduces micronutrient deficiencies, and improves iron uptake (WFP Rwanda 2019).

Freeing up resources to allocate more of them to the meals delivered rather than to the logistics and transportation involved has tangible economic benefits over the long term at both the individual and the national levels. Improving the transport supply chain that the HGSF program relies on can help mitigate the impact of shocks and prevent acute and chronic food and nutrition insecurity in children, one of the most vulnerable populations in any society. In Rwanda, child mortality associated with malnutrition has already reduced the country's workforce by 9 percent (UNICEF Rwanda, n.d.). Interventions that help reduce malnutrition are thus associated with enormous benefits for the quality of life of the population as well as for the long-term economic growth of the

East Africa region. A cost-benefit analysis conducted in 2017 concluded that every dollar invested in school meals in Rwanda can generate a return of $4.80 and $5.60 for home-grown and in-kind modalities, respectively, over a child's lifetime (MasterCard 2017). The WFP estimates that there is a $10 overall return for a $1 investment in school meals, measured by education, health, and productivity gains.[9]

NOTES

1. U.S. Department of Agriculture and IndexMundi. Note that maize and millet/sorghum consumption figures are not available for Burundi, Djibouti, or South Sudan, but are imputed by assuming that per capita consumption equals the average of those countries for which the data were available.
2. *Burundi, Kenya, Rwanda, Tanzania, Uganda; South Sudan since 2016.*
3. These changes include supply chain optimization, which reduced the number of required airlifts, and changes in the local demand for and availability of food aid.
4. Actual volumes airdropped are likely proportional to village population; however, population data are not available.
5. This implies an average overall flight duration of as much as 2 hours and 45 minutes for a distance of 400 km (despite theoretical cruising speed of the cargo plane of at least 600 km/h); speed is low on takeoff, climb, approach to drop zone, during airdrop, and approach to airport.
6. Although the stated payload capacity of the Flyox is 1.8 mt, there are limitations in the cargo compartment volume that effectively reduce the actual amount of cargo that can be carried.
7. The cost of financing is not included.
8. Brochure on the Flyox I, http://singularaircraft.com/pdf/cataleg_SA_en_MAY2016 _LR.pdf.
9. "School Feeding: Launch of the State of School Feeding Worldwide 2020 report," https://www.wfp.org/school-feeding.

REFERENCES

Bashir, S., M. Lockheed, E. Ninan, and J.-P. Tan. 2018. *Facing Forward: Schooling for Learning in Africa.* Washington, DC: World Bank.

Cervigni, R., A. Losos, P. Chinowsky, and J. E. Neumann, eds. 2017. *Enhancing the Climate Resilience of Africa's Infrastructure: The Roads and Bridges Sector.* World Bank, Washington, DC.

China Daily. 2018. "JD's First Large-Scale UAV Debuts in Shaanxi." November 20, 2018. https://www.chinadaily.com.cn/a/201811/20/WS5bf3af9ca310eff303289e31.html.

De Serio, C. J. 2019. "Disclosable Version of the ISR–Lake Victoria Transport Program–SOP1, RWANDA–P160488–Sequence No: 06." World Bank Group, Washington, DC.

Gadhia, K. 2017. "Lake Victoria Challenge Heavy Lift Session." Presentation at Astral Aerial: Lake Victoria Challenge, Mwanza, Tanzania, October 31.

Hoeben, J. S. F. 2014, "A Value Analysis of Unmanned Aircraft Operations for the Transport of High Time Value Cargo." Delft University of Technology, Delft.

INFORM. 2015. Index for Risk Management. Inter-Agency Standing Committee, European Commission. Data accessed through Humanitarian Data Exchange.

ITF (International Transport Forum). 2018. "(Un)certain Skies? Drones in the World of Tomorrow." International Transport Forum, OECD, Paris.

MasterCard. 2017. "The School Feeding Investment Case: Cost-Benefit Analysis in Rwanda." Report prepared by MasterCard for World Food Programme.

NISR (National Institute of Statistics of Rwanda). 2015. "Rwanda Poverty Profile Report 2013/2014: Results of Integrated Household Living Conditions Survey [EICV]." National Institute of Statistics of Rwanda, Kigali.

Rupani, S. 2017. "UAVs in Supply Chains: Thinking about UAVs as a Transport Mode." LLamasoft, Inc., Global Impact Team.

Snilstveit, B., J. Stevenson, D. Phillips, M. Vojtkova, E. Gallagher, T. Schmidt, H. Jobse, M. Geelen, M. G. Pastorello, and J. Eyers. 2015. *Interventions for Improving Learning Outcomes and Access to Education in Low- and Middle-Income Countries: A Systematic Review.* 3ie Systematic Review 24. London: International Initiative for Impact Evaluation (3ie).

Soesilo, D., P. Meier, A. Lessard-Fontaine, J. Du Plessis, and C. Stuhlberger. 2016. "Drones in Humanitarian Action: A Guide to the Use of Airborne Systems in Humanitarian Crises." FSD (Swiss Foundation for Mine Action), Geneva.

Taban, E. 2019. "Disclosable Version of the ISR–Rwanda Feeder Roads Development Project–P126498–Sequence No: 13." World Bank Group Washington, DC.

UNICEF. 2017. "Tanzania Nutrition Fact Sheet." https://www.unicef.org/tanzania/media/751/file/UNICEF-Tanzania-2017-Nutrition-fact-sheet.pdf.

UNICEF. 2019. "Food and Nutrition Situation." South Sudan Country Office, October. https://www.unicef.org/southsudan/media/2091/file/UNICEF-South-Sudan-Nutrition-Briefing-Note-Aug-2019.pdf.

UNICEF Ethiopia. 2019. "Humanitarian Action for Children." UNICEF Ethiopia. https://reliefweb.int/report/ethiopia/humanitarian-action-children-2019-ethiopia.

UNICEF Rwanda. not dated. "Nutrition." UNICEF Rwanda. https://www.unicef.org/rwanda/nutrition#:~:text=Child%20mortality%20associated%20with%20malnutrition,cent%20to%2038%20per%20cent.

Van Groningen, R. 2017. "Cost Benefit Analysis Unmanned Cargo Aircraft: Case Study Stuttgart – Urumqi/Shenzen." Erasmus University Rotterdam, Rotterdam.

WFP (World Food Programme). 2018. "Rwanda Country Strategic Plan (2019–2023)." World Food Programme, Rome.

WFP (World Food Programme) Rwanda. 2019. "Decentralized Evaluation: WFP's USDA McGovern-Dole International Food for Education and Child Nutrition Program's Support in Rwanda 2016-2020 Evaluation Report: Midterm Evaluation." World Food Programme Rwanda, Kigali.

World Bank. 2018. Lake Tanganyika Transport Program – SOP1 Tanzania. PID/ ISDS." World Bank, Washington, DC.

3 Land Mapping and Risk Assessment

THE BIG PICTURE: THE MARKET AND THE NEED FOR AERIAL MAPPING AND RISK ASSESSMENT

The United Nations lists land tenure as one of the main preconditions for alleviating poverty by 2030, the first Sustainable Development Goal. The lack of accurate maps more broadly is a significant problem in parts of Africa for disaster relief agencies, local authorities, and people looking for safe places to build homes. Could inexpensive survey drones and local volunteers help fill the gaps? Mapping, both commercially for companies or governments and for local communities, is a vital function of drones in Africa, where only 3 percent of the land is mapped to local scale compared with 90 percent in Europe. Only 10 percent of Africa's rural land is registered; the remainder is undocumented and informally administered, which makes it susceptible to land grabbing and expropriation without fair compensation. In Uganda alone, current cadastral government authorities estimate it will take the few dozen surveyors in Uganda 1,000 years to legally register the approximately 15 million unregistered land parcels in the country (Wickless and Westers 2020). Similarly, there are only 1.9 land surveyors per 10,000 square kilometers (km²) in Tanzania, compared with 41 in the United States.[1] Informal settlements housing large numbers of urban residents are expected to expand disproportionately around the world through rural-urban migration over the next decades (Bayle et al. 2020). In addition, recognition of land rights and titles is perceived as increasingly important in the region given that land can serve as the most readily available form of collateral in loan transactions.

Demand for disaster risk mapping and postdisaster risk recovery planning is growing in the region with the increased frequency and unpredictability of various natural disasters. Base mapping of urban areas needs to be updated frequently with the rapid growth of the region's cities. Postdisaster assessments at the country level, similarly, are needed nearly every year. Many quickly growing or expanding medium to small municipalities and communities lack the resources with which to monitor the expansion of informal settlements, to conduct and update assessments of exposure and risk, and to plan interventions to increase resilience (Bayle et al. 2020).

AVAILABLE NONDRONE-BASED APPROACHES

One of the most significant hindrances to the formalization of land rights in many countries is the rigidity of conventional cadastral surveying regimes. Attempts to improve efficiency by the introduction of new approaches do not always succeed if they cannot convincingly demonstrate compliance with existing standards and procedures (Volkmann 2020). Until recently, mapping and survey data were acquired either via surveying personnel on the ground, traditional manned aircraft, or satellite imagery. Although the benefits of modern land administration systems are clear, the process for developing such systems is complex, with many projects spanning a decade or more before delivery. Adding to the complexity is the availability of and risks to human resources, particularly surveyors (Smith and Orçan 2020).

Traditional airborne surveying methods often require high monetary outlays and imported skills and services that may not be feasible in the African context. In a 2019 risk-mapping activity funded by the World Bank in West Africa, the cost of using traditional aircraft to conduct an aerial survey of an area of 450 km² was an estimated \$130,000, or about \$290 per km². Equipment costs accounted for about 30 percent of the total, and data analysis was the second most costly input, accounting for 14 percent of the total, according to documentation provided by the World Bank team managing the activity.

Traditional airborne survey methods may also be associated with various practical challenges such as weather conditions. A 2015–16 mapping project undertaken by a consortium of Pasco Japan and COWI Denmark, aimed at mapping the full 224,000 km² area of Uganda using aerial cameras, with the goal of making a base map for major development programs in Uganda with a focus on land administration, demonstrates the advantages and challenges of the traditional manned aircraft approach (Noergaard, Kristensen, and Sato 2020). The mapping effort involved the creation of high-resolution (15 centimeter [cm] per pixel) orthophotos[2] of the 50 largest cities (4,000 km²) and medium-resolution (40 cm per pixel) orthophotos for the rest of the country, with a network of more than 600 physical ground control points (GCPs) established in the field to ensure an accurate geographical reference. Given that the aerial data collection was carried out with turboprop aircraft flying at an altitude of at least 6,000 meters, a major challenge for the project was weather because parts of the country are rainy, and cloud coverage makes it difficult to obtain cloud-free images at that altitude. Another major challenge was the presence of relatively few airports, some with limited availability of jet fuel for the turboprop aircraft. Finally, establishing GCPs in a regular grid was difficult in hard-to-reach parts of the country. The overall task therefore required a considerable amount of time: three June–September seasons. Reduction in the number of GCPs was considered an important cost- and time-saving measure for future projects of this type.

New technological developments, such as single photon LiDAR (SPL; LiDAR = light detection and ranging), may provide a significant advantage in thin cloud and ground fog conditions and improve image collection efficiency by three to five times. In the past five years, the SPL technique has been used in several large-scale (up to 450,000 km²) projects; in a project in Navarro, Spain, 10,000 km² were covered in only 50 hours' effective flying time (albeit spanning over three months) using a 9-seat turboprop aircraft flying at a height of 3,900–6,000 meters (Roth and Zalba 2020).

Another currently available alternative to drone-based solutions is free aerial imagery provided by companies such as Google Earth and Landsat, although many rural areas in Africa are poorly covered. These solutions have been available for many years; however, the resolution of the imagery varies greatly across the world (Li 2020). In these contexts, open-source, collaborative mapping platforms such as OpenStreetMap can provide a partial solution, enabling volunteers to create maps for a remote region by contributing ground features. So-called citizen science methods, also known as crowdsourcing, human computation, or volunteered geographic information (VGI), make use of free multispectral band imagery from platforms such as Google Earth and are considered competitive with traditional survey approaches from the cost-minimization perspective. These novel approaches have been used to classify the informal sector and other urban land use types in particular. User contributions in VGI projects involve tracing out specific land use types on *georeferenced images; however, in some cases, they may also entail the collection of global positioning system (GPS) points by the users* (Arsanjani and Vaz 2015). The citizen science approach may be superior to other land use classification approaches (such as machine learning), especially where features are heterogeneous and the classification typology is inconsistent (Albuquerque, Herfort, and Eckle 2016). However, the results of this approach from the perspective of the precision of land cover maps or the cadastral databases created have been variable (see Chingozha et al. [2020] for a pilot initiative in Zimbabwe using student volunteers).

Specialized mobile applications are becoming available that connect para-surveyors (local representatives in the community) and national land databases. These applications claim to be a cost-effective solution for improving cadastral data quality in urban areas (Aditya et al. 2020). For example, in Uganda, where registered land accounts for less than 20 percent of plots, a mobile application is being piloted in the Buliisa District by the Ministry of Lands, Housing and Urban Development and the Cadasta Foundation, a nonprofit focused on sustainable and fit-for-purpose approaches to securing land rights, to assess the feasibility of district and subcounty-level institutions locally and cost-effectively capturing and integrating customary land[3] into the newly created National Land Information System (Sanjines Mancilla et al. 2020). Similarly, artificial intelligence–based models have been developed to enable the automatic detection of the boundaries of distinct plots encompassed within a satellite image (Raut et al. 2020).

In risk assessment and postdisaster response, new tools are being deployed that bring together the emerging capabilities of frequent and free high-resolution satellite imagery, artificial intelligence, and local knowledge. These tools integrate analysis of a large variety of data such as satellite imagery and local precipitation and wind data, and claim to be able to provide rapid and low-cost identification of exposure to climate-related risks, shocks, and stresses for informal settlements more economically, quickly, frequently, and transparently than current approaches (Bayle et al. 2020). The World Bank and the German Aerospace Center (Deutsches Zentrum für Luft- und Raumfahrt, DLR) are working on a joint project to improve mapping of the spatial distribution and pace of urban exposure growth to provide a better understanding of risk in African cities. The work will bring together the emerging capabilities of frequent and free high-resolution satellite imagery, machine learning techniques, and local knowledge (including data from official sources as well as VGI and social media) to obtain a more accurate picture of how many people in African cities are exposed to natural hazards and where this number is growing, thus supporting the prioritization and distribution of urban resilience resources and investment (Esch et al. 2020).

INTEGRATING DRONES INTO EXISTING MAPPING AND RISK ASSESSMENT SYSTEMS

The cases in which unmanned aerial vehicles can add value

Drones provide a practical advantage in cases in which a smaller area of interest needs to be covered, the level of detail needed is higher, and cloud cover presents challenges to satellites and airplanes (World Bank and Humanitarian OpenStreetMap Team 2019). Drones can also provide an advantage in physically challenging environments (see box 3.1). According to in-depth analysis of the use of drones in humanitarian crises, with an emphasis on natural disasters (Soesilo et al. 2016), the use of drones for mapping presents significant advantages compared with other methods in the following cases:

- Capturing aerial imagery and making base maps of relatively small areas (less than 15 km²)
- Collecting optical imagery where cloud cover precludes the use of satellites and airplanes

BOX 3.1

Adoption of drones for land mapping in physically challenging environments: Nepal

In Nepal, nonstatutory or informal land tenure is estimated to apply to about 25 percent of the total arable land and settlements, or about 10 million parcels. Therefore, a significant number of Nepalese live in informality, without any legal recognition or spatial recordation, and, consequently, without security of tenure. Identification of and training on tools that can help support the delivery of the country's land policy goals are now needed.

Unmanned aerial vehicles (UAVs) are emerging as an option that could be especially suitable for remote, difficult-to-access areas. Nepal's Department of Survey has procured hybrid drones and is working on setting up a drone mapping lab for cadastral, topographical, and boundary mapping. In addition, unlike in other countries, it is relatively easy to get flight permission from local authorities for lightweight UAVs (less than 2 kg) for surveying and mapping purposes. Moreover, the small size of the tech sector in Nepal means that collaboration between the government and the private sector is well established. Finance is not considered a barrier to using UAVs, and the technology and its ability to produce high-quality imagery in remote and rural areas is seen as a major advantage. Educational institutions for survey and mapping are already training on the tools and techniques. However, although there is a growing body of UAV-derived imagery, the coordination and sharing of that data happens only on a piecemeal basis, if at all. Similarly, expectations of the likely UAV impacts in the field of land administration may need tempering, given that the technology so far has not significantly increased the number of parcels mapped and registered.

Source: Dijkstra et al. 2020.

- Operating in dense and fast-changing environments such as urban areas and refugee camps
- Creating accurate elevation models needed for flood, avalanche, and debris flow modeling and rubble volume calculations
- Making 3D renderings of buildings and geographic features

Depending on the level of detail required, typical altitudes for a mapping drone may range between 80 and 300 meters. The drone-mapping workflow consists of, first, taking a series of images along a preprogrammed flight path above the mapping area; second, downloading the images from the drone camera and stitching them together using mapping software; and, finally, producing the final, geometrically correct aerial image, or orthomosaic (Soesilo et al. 2016).

The unmanned aerial vehicle (UAV) technology for mapping is relatively well understood, and a number of sensors that can create safe and controlled flights are readily available. Fixed-wing drones need considerably more space for take-off and landing compared with multirotors. However, because of their improved flight dynamics they can cover much larger areas, making them ideal for applications such as land surveys. Traditional GPS can be used to achieve relative accuracy of within a few meters. Variations such as real-time kinematic correction or postprocessed kinematic correction make possible absolute precision at the centimeter level. Accelerometers, gyroscopes, and barometers can all be integrated into UAVs to provide data to the flight control systems, thus achieving controlled, predictable, safe flight.

Aerial mapping is perhaps the most developed use case for drone technologies in East Africa. The Humanitarian UAV Network and WeRobotics are developing local flying labs for personnel training purposes (USAID 2017). A network of drone labs—Flying Labs—has been established throughout the African continent. Flying Labs has conducted several mapping projects across the East Africa region, combining drone technology with satellite imagery and verification. In Ng'hoboko, a sparsely populated farming village in northern Tanzania, land registries are almost unheard of, and property is handed over from generation to generation. An accessible digital record of land and resource rights information can empower local communities to make data-driven decisions resulting in better economic outcomes. With funding from the Hewlett Foundation, Flying Labs Tanzania used drone-gathered data to create digital elevation models for land use plans that could be used as a basis for issuing certificates of customary right of occupancy or traditional title deeds to farmers. However, the drone technology served only as an input into the overall land surveying process; ground verification (walking the boundaries of each farm with a handheld GPS device accurate to plus or minus three meters) and feedback from the farmers themselves were essential additional inputs. Participatory mapping using the drone-based high-resolution maps was organized in collaboration with the Economic and Social Research Foundation about four to five months after the drone image collection to obtain feedback from the eventual users of the mapping exercise outputs. The close collaboration with community elders was indispensable because they provided not only technical and logistical support but also lent the project credibility that inspired trust among the farmers and ensured the safety of the staff involved in the project while navigating the land.

Drones in the region are also used relatively widely in disaster risk assessment, postdisaster damage assessment, and postdisaster recovery planning. The different components of a robust emergency preparedness and early warning system can benefit from georeferenced aerial data that are acquired with the help of drones; overlaying existing geographical data on the newly acquired aerial images can provide a much better overview of potential evacuation and supply needs, as well as demands for sanitation, education, food, and health care (UNICEF Malawi 2019). The contribution of drones to the preparation of disaster management plans via hazard mapping is among the most compelling use cases. In Malawi, for example, UNICEF is using drones to map flood-prone areas, collecting data to create "disaster profiles" within districts. The results of the drone imagery data analyses are then used by the government at national, district, and community levels to make more informed decisions based on the risks associated with residing in flood-prone areas (UNICEF Malawi 2018). In Tanzania, drone imagery was used to derive detailed drainage maps that can assist in flood risk reduction planning (Lawani et al. 2017).

After a disaster, the ability to rapidly acquire and analyze aerial imagery becomes very important, and drones can help quickly estimate the damage to buildings, to infrastructure, or to crops, as well as detect survivors. In these situations, time is of the essence. In Malawi in 2017, UAVs were used in Karonga and Salima to assess the extent of damage caused by flooding, paving the way for inclusion of the technology in the government's emergency response and preparedness strategy. In 2019, UAVs were used to assess the impact of devastating floods throughout Malawi's Southern Region. The data acquired in both 2017 and 2019 are being used to develop flood preparedness models that will assist in ensuring the well-being of communities affected by recurring floods.

Drones are currently used to perform visual impact assessments of natural disasters such as floods and cyclones and to assess access to and prioritization of areas (ITF 2018). In 2016, UNICEF supported the introduction of drones in aerial imagery collection by the National Office for Risk and Disaster Management in Madagascar. In 2017, a humanitarian drone corridor was launched in Kasungu District in Malawi to pilot the use of drones to acquire imagery and monitor flooding during an emergency response because a combination of severe droughts and floods had led to severe food shortages in the country in 2015 and 2016. Using drones, in less than 24 hours the government was able to gather real-time high-altitude photos of flooding conditions and infrastructure damage in two districts. In preparation for the next rainy season, UNICEF supported training of personnel from government, United Nations agencies, and nongovernmental organizations to integrate drones into an emergency response plan, using a simulation exercise (UNICEF 2017). With World Bank support, drone imagery has been used in Caribbean nations such as Sint Maarten to support posthurricane planning and reconstruction efforts (see box 3.2).

An additional consideration in favor of UAVs is the ability to produce high-resolution, ortho-rectified mosaics, which requires specific flight plans and altitudes that are typically not achievable with helicopters. In the aftermath of Cyclone Pam in Vanuatu in 2015, which affected 132,000 people, the World Bank chose to use UAVs to carry out aerial surveys of the disaster-affected areas because of the limited and unpredictable availability and cost of chartered helicopters in Vanuatu (Soesilo et al. 2016). However, the presence of local drone capacity was deemed necessary for smaller storms that do not attract as much international donor funding to allow rapid damage assessments to be conducted affordably.

BOX 3.2

Global Program for Resilient Housing: Collecting drone imagery in the Caribbean

Caribbean nations are increasingly being hit by the strong winds and intense rainfall of hurricanes. The human and economic toll of recent category 5 storms such as Dorian (2019), Irma (2017), and Maria (2017) left island states crippled with debt and caused a growing interest in measures that could reduce housing vulnerability.

To support recovery and reconstruction efforts, the World Bank's Global Program for Resilient Housing (GPRH) developed a methodology that generates a detailed building database. The team leverages drone images and machine learning algorithms to characterize roof condition and material. Street view imagery then complements the rooftop analysis with information about the quality and use of each building (residential, commercial, critical infrastructure, and so on).

After Hurricane Irma, GPRH worked with the government of Sint Maarten to assess the island's reconstruction efforts. Faced with extensive damage and limited housing data with which to monitor the reconstruction effort, the government was interested in capturing images. Drone images were collected over 28 km² (4–6 cm resolution; red, green, blue), and more than 13,000 rooftops were evaluated (photograph B3.2.1). In addition to rooftop analysis, high-risk neighborhoods such as informal settlements near a new trash dump were studied. During the recovery, backyard homes sprang up, and drones allowed them to be identified and counted. Finally, the drone-derived information on the number, size, and quality of homes provided actionable information to different stakeholders, including government officials, as they assessed previous efforts and planned new investments.

Caribbean countries are ideal for drone application, being small with fast-changing weather. Drones can quickly capture imagery over islands and are cost-effective compared with airplanes or field surveys. With the benefit of drone technology, GPRH offers detailed housing databases to identify opportunities to expand, strengthen, and protect existing homes, supporting Sint Maarten's effort to build better before the next hurricane.

PHOTOGRAPH B3.2.1

Roof delineation using drone imagery

Source: © Global Program for Resilient Housing / World Bank. Permission required for reuse.

To ensure that drones in the immediate aftermath of a disaster are used effectively and promptly, organizations can build local or regional capacity and integrate them into their emergency response toolkits (Soesilo et al. 2016). The most successful use of drones for mapping in humanitarian emergency contexts usually involves an organization that already possesses the necessary equipment, authorization to fly, and skilled human resources at the time of the disaster. However, only a few organizations have invested in such in-house capacity, and the majority work with external service providers for drone deployments. In Malawi, the Department of Civil Aviation has been working to integrate drones into flood emergency response, the monitoring of infrastructure programs, and the overall national disaster preparedness and response framework (Chadza 2019).

Still, drone-based mapping in the East Africa region remains relatively poorly integrated into overall government systems; mapping assignments are often fully contracted out to private sector firms that deliver the finished product using their own staff and systems, with little involvement of local authorities and minimal transfer of knowledge and expertise. The interviewed drone operators questioned whether, despite the relative maturity of drone technology for mapping purposes compared with other applications, this solution can feasibly be integrated into the mainstream work of national land ministries. World Bank staff working on disaster risk–related drone applications in the region highlight drone technology maintenance needs, which typically cannot be fully met by local governments and, instead, make working through local specialized contractors more practical.

For large mapping initiatives, the experience in East Africa shows the importance of close coordination between the international research and intergovernmental organizations that design and financially support these initiatives and the local academic and government institutions, with the intention of long-term knowledge transfer and continuity. Since its inception in 2016, the Zanzibar Mapping Initiative has become representative of how African nations can approach the urgent geospatial challenges they face. Critical to the success of the Zanzibar Mapping Initiative, as a result of which 80 percent of the island's land was mapped, has been the State University of Zanzibar's unmanned aerial system mapping program, which is helping the government of Zanzibar's Commission for Lands map the island using low-cost drones (UbuntuNet Alliance 2018).

Both the Zanzibar and the Dar es Salaam drone-based mapping projects have clearly demonstrated that an aerial photo can only show so much. Personnel on the ground are also needed to label what the drones see and flag issues, such as blocked drains. Creating a map takes a great deal of skill, and even the largest global technology companies say they need local help to complete their maps. Many global mapping companies may find insufficient commercial incentive to cover Africa, which is why governments, local communities, and open-source data are important (BBC News 2019).

Potential time and cost savings and quality improvements

Land mapping

Satellite imagery used in mapping and risk assessment has traditionally been able to provide imagery for areas as large as 10,000 km^2 per day; however, it may be less effective in the tropics because of cloud cover, and the level of detail provided will typically not exceed 30–50 cm per pixel compared with 3–10 cm per pixel for most UAVs (World Bank and Humanitarian

OpenStreetMap Team 2019). For example, fetched satellite image data may also fail to seamlessly detect boundaries between individual land plots via state-of-the-art approaches in instances in which boundaries between the plots are occluded by trees or are made less legible by nonuniform and abrupt crop growing patterns (Raut et al. 2020). Using manned aircraft is another alternative, but not feasible if needed every six months because of the high cost—just mobilizing the plane can cost hundreds of thousands of dollars (BBC News 2019).

An important advantage of UAVs in area assessments, as compared with traditional manned-aircraft platforms or satellites, lies in their agility in collecting aerial imagery. In addition to their flexibility and speed, UAVs are also able to more easily access hard-to-reach locations (ITF 2018). Commercial UAVs are small and ultra-light, facilitating an affordable mapping service through a process that takes days or weeks, rather than months or years, from planning to product. However, perhaps the most promising aspect of UAV technology is the reduction in capital and skills required for data collection and the ability to rapidly deploy to the field. There is thus a real possibility that geospatial information can be gathered and processed at the local level by local community members and on short notice. The independence from highly centralized and remote institutions for actionable, time-sensitive information may bring about significant improvements in some domains, including agricultural production and customary land tenure (Volkmann 2017).

However, the rich streams of data generated by drones are only useful if they can be stored and analyzed (Mbuya, Rambaldi, and Chaham, n.d.). Data storage facilities, analysis software and the computing power to run it, fast internet connectivity where analysis is done in a cloud computing environment, and accessible power supply are all components of the technology. So, too, is a workforce of flight planners, pilots, analysts, and advisers. However, affordable or even free solutions are coming onstream for drone data processing. For example, OpenDroneMap is an open-source toolkit for processing civilian drone imagery; appropriate for highly overlapping unreferenced imagery, the software permits unstructured data (simple photos) to be turned into structured geographic data such as colorized point clouds, digital surface models, textured digital surface models, and highly detailed orthophotography.[4] Similarly, free, high-resolution, and up-to-date satellite imagery, which is necessary for efficient UAV flight planning, is available across most of the African continent. As more countries introduce UAV regulations that require pilot licensing, and more training institutes are established to award these licenses, many young Africans are likely to take up the opportunity to start careers as licensed drone pilots.

The up-front cost for mapping drones can range from about $1,000 to $100,000, depending on range and other technical parameters that also affect the area that can be covered in a specific period. For example, up to 5 km[2] can be covered in one day under standard conditions[5] using the Mavic Pro drone ($1,000), but the more expensive eBee drone ($15,000) can cover five times the area within the same time.

Several factors typically influence the required budget in drone deployment for mapping purposes (Soesilo et al. 2016):

- *Preparation.* Applying for licenses, negotiations with ministries or competent authorities, and engagement with local communities
- *Data collection.* Preparing flight plans, piloting and equipment maintenance

- *Data processing.* Uploading, processing, and rendering of the collected data to create orthomosaics, base maps, and 3D models where access to power, internet, and processing platforms may be restricted
- *Data analysis.* Obtaining actionable information from data analyzed by specialists

Depending on travel distance to the survey area, flying the drone and collecting imagery usually accounts for less than half of the total costs of a mapping project, as noted by the interviewed drone-mapping experts at the World Bank. Each flight can yield several hundred images that need to be stitched with very expensive software and hardware. And, while free and open-source software options are becoming available, most drone mappers continue to rely on commercial photogrammetric software. Another significant cost in addition to flying and processing the data is the placement and surveying of GCPs that are required for generating professional products for platforms that are not equipped with high-accuracy positional sensors.

Larger areas or higher resolution not only require more time for image collection but also increase the amount of data that is produced, which in turn increases the need for storage capacity, processing capacity, and processing time, all of which can increase costs. According to the interviewed drone operators involved in mapping projects in East Africa, the economic competitiveness of drones compared with other surveying methods depends on the size of the land area to be assessed: as land area increases, the operating costs of drones tend to increase at a faster rate compared with those of satellite imagery analysis–based and aircraft survey–based methods (figure 3.1). This relationship is consistent with the World Bank and Humanitarian OpenStreetMap Team assessment (2019) that suggests that drones will present a cost advantage compared with satellites and airplanes if small areas (less than 10 km²) need to be assessed but will be much more expensive than the alternative approaches if the area of interest exceeds hundreds of thousands of square kilometers.

The cost-effectiveness of drones compared with manned aircraft also depends on the availability of local surveying companies and the type of products needed. For example, the high-precision LiDAR sensor-based surveys

FIGURE 3.1

Schematic relationship between covered area and cost for alternative approaches of image capture

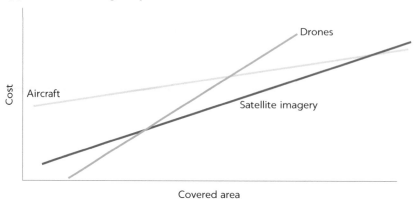

Source: Interviews with East African drone operating companies.

almost always require a manned-aircraft setup for areas larger than a few square kilometers because of the larger batteries that need to be operated (and carried), whereas small fixed-wing drones such as the eBee are likely still more cost-effective for generating orthomosaic imagery for areas up to 100 km².

In the Ng'hoboko land-mapping project in northern Tanzania, Flying Labs attempted to gain a better understanding of the realistic costs of drone data collection and the additional value that drone data can contribute, and to identify the remaining challenges in acquiring drone data for land management applications and in combining it with other data sources. The project mapped an area of 12,000 hectares in the Meatu District in which Ng'hoboko (3,500 hectares) is located, which the drone company staff interviewed for the current study considered the upper limit to be surveyed by drone technology. For the assignment, two Swiss-made, fixed-wing, survey-grade mapping drones with excellent flight controllers and the ability to fly beyond visual line of sight at a cost of approximately $20,000 were used, *involving two pilots who were paid $200–$300 per day.* No locally produced drones were found to be able to perform at the level required in this exercise. As reported by the staff involved in the project, the ground verification and participatory mapping that followed the drone-based data collection and image processing involved yet additional, nonnegligible operational costs (teams went back with printed copies of the aerial maps to work with communities on issues ranging from flood planning to land usage). However, this aspect of the project was essential to the overall acceptability and credibility of the mapping results. It took five days to process the Ng'hoboko-specific imagery using a powerful and relatively automated processing machine (more than 32 gigabyte random-access memory). As a rule of thumb, one day of drone-based image acquisition is typically associated with three days of image analysis in assignments of this type. Data analysis was identified as the most important aspect of the overall project in which drones can serve as a tool for data acquisition.

According to experts, improved productivity is the main benefit of using drone technology compared with land surveyors or similar methods. Thus, for a village the size of Ng'hoboko, drone-based surveys are the appropriate approach. One week's worth of work using ground surveys was accomplished by drones in less than a day. The drone-based assessment in Ng'hoboko, which produced images with resolution of 50 cm per pixel, enabled all farms in the study perimeter to be identified and nine out of ten of them were correctly traced. The level of accuracy achieved was considered impressive despite the semi-arid nature of the environment at the time the data were collected.[6] Overall, the staff involved in the project indicated that although drone data can accelerate the land assessment and titling process, it is not a magic bullet. Consolidation and digitization of forms and extensive community engagement are also necessary to achieve positive results.

In 2016, the Ministry of Lands, Housing and Human Settlements Development in Tanzania conducted a project to evaluate unmanned aerial systems as an option for acquiring aerial imagery to assist the national land tenure support program. On April 1, 2016, the program asked the Tanzania Commission for Science and Technology (COSTECH) for use of their drones for the purpose of aerial mapping of pilot sites for the program (COSTECH 2016). COSTECH agreed to provide its support in collaboration with the World Bank and Uhurulabs. The mission was conducted over two weeks; its primary objective was to perform an aerial survey of two areas in Kilombero and Ulanga Districts, each about 9 km², to generate an orthomosaic with a ground sample distance (GSD)[7] of 10 cm. A secondary objective was to confirm the accuracy of the drone data with and without GCPs.

The primary objective was achieved; COSTECH was able to deliver data for almost 800 percent more area (147 km² rather than 18 km²) and a higher resolution than what was requested (GSD of 7 cm rather than 10 cm). The imagery obtained was of sufficient resolution to easily identify and mark plots of land and other features. The absolute accuracy without GCPs was 4.1 meters, which is within the expected range, thus achieving the project's secondary objective. The Ministry of Lands could confirm absolute accuracy of within 2 cm.

In Rwanda, a group of researchers in collaboration with landowners conducted drone flights to map specific areas in the country to assess whether UAVs could be used to make orthophotos. Nearly 1,000 images were collected using a low-cost DJI Phantom 2 Vision+ quadcopter (retail price of about $1,000) that flew autonomously at an altitude of 50 meters above the ground. An orthophoto covering 0.095 km² with a spatial resolution of 3.3 cm was produced and used to extract features with subdecimeter accuracy. Quantitative assessments suggested that the orthophoto-replicated field measurements were within 0.6 percent of the actual dimensions. The duration of the flight over the study area, including takeoff and landing, was approximately two hours, and identification and marking of the GCPs in the images and image orientation, dense image matching, and orthophoto creation took about two days on a good-quality desktop computer (Koeva et al. 2016).

As reported by the local drone operators in Tanzania interviewed for this report, the estimated cost of a drone-based mapping task of the type led by COSTECH in 2016 was approximately $575 per km² mapped. The largest share of the cost, about 57 percent, is for the data acquisition itself; however, as much as 39–40 percent is allocated to capacity building. The remainder, about 3 percent, is associated with data processing. Figure 3.2 shows a more detailed cost breakdown. The data acquisition figure does not include the initial purchase price for the UAVs or the ground equipment (ground control station, processing machine, and processing software), which the project client should purchase so that it is available for future use. The estimated cost of these initial capital investments amounts to an additional $39,500, of which about three-fourths is the cost of the UAV. With regard to the time required to do the mapping work, the local drone operators report that a target area of 20 km² can be covered over the course of a single day with three UAV flights; however, the overall mission duration is about three days.

Risk mapping and postdisaster assessment

In disaster risk assessment and postdisaster recovery, drones can provide significant time savings compared with traditional assessment methods, assuming they can be deployed quickly. For example, in 2017, drones were used in a disaster management simulation in Malawi and it took just hours to process the data. The following year, new software cut the processing time by more than half.[8] In addition, drone technology also offers potential cost advantages compared with other risk assessment and postdisaster planning approaches, although the savings depend on scale: airplane-based assessments may still be competitive; however, such assessments involve specific logistics issues. Compared with traditional aircraft, drones may also be more difficult to deploy in urban environments (considering the need for interaction with air-control systems on the ground).

FIGURE 3.2

Cost breakdown of a UAV-based land-mapping task

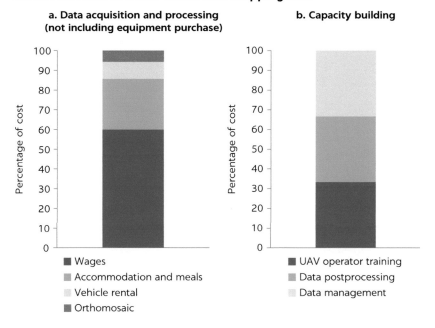

a. Data acquisition and processing
(not including equipment purchase)

b. Capacity building

- ■ Wages
- ■ Accommodation and meals
- □ Vehicle rental
- ■ Orthomosaic

- ■ UAV operator training
- ■ Data postprocessing
- □ Data management

Source: Interviews for this study with local drone operators in Tanzania.
Note: UAV = unmanned aerial vehicle.

Compared with satellite image–based risk and postdisaster damage assessments, drones may involve more permitting requirements and be associated with higher costs; however, they obviate licensing issues that are common to some satellite image platforms used to create digital elevation models that are used in risk assessment. In general, satellite imagery provides lower resolution (as illustrated in photograph 3.1) but covers larger areas than drones. Drones also have the advantage of enabling aerial surveys of the same area but with higher frequency, which is important in the monitoring of disaster events and which would not be possible with traditional aerial surveys. As noted, the drone's ability to fly below the clouds to capture images may sometimes provide a significant advantage, given that heavy cloud cover is often present for several days following extreme weather events such as typhoons, cyclones, and hurricanes (Soesilo et al. 2016).

With World Bank and Global Facility for Disaster Reduction and Recovery assistance, drone technology was used in the assessment of postdisaster damage in eastern Uganda following a deadly landslide in Bududa District in October 2018. The mudflow that was caused by the landslide affected 13 villages and 12,000 people in an area stretching more than 10 km up a steep hill into Mount Elgon Forest, which is difficult to access from the ground. Covering an area of interest of about 5 km^2, the postdisaster drone-based aerial surveys were contracted out to local and regional drone operation companies. Specific activities included identification of the affected area (using satellite imagery and other data) and creation of flight plans for UAV surveys, UAV surveys to generate detailed imagery and terrain data sets, and creation of an orthomosaic (3–7 cm GSD) and a digital terrain model (less than 15 cm GSD) of the affected area (World Bank and GFDRR 2018). The timeline for the work was less than two weeks at an estimated overall cost of $16,500, or $3,300 per km^2.

Comparison of satellite imagery and drone imagery resolution

a. Satellite imagery: DigitalGlobe Standard   b. Drone imagery: CC-BY 4.0 Open Imagery Network

Satellite imagery - DigitalGlobe-Standard   Drone imagery, 2017-05-23 - CC-BY 4.0 Open Imagery Network (danbjoseph)

Source: "What Is This All About?" © Dan Joseph / https://danbjoseph.github.io/drones/. Used with the permission of Dan Joseph / https://danbjoseph .github.io/drones/. Further permission required for reuse.

In another World Bank–funded activity, in the islands of Fiji and Tonga, where the conditions and scale of imagery needed may be comparable to the islands in East Africa's Great Lakes,[9] collection of aerial imagery with the objective of creating 2D orthomosaics for risk assessment and postdisaster recovery planning ranged in cost between about $39,000 for an assignment covering a 10 km² area ($3,900 per km²) to about $57,600 for an assignment covering a 50 km² area ($1,150 per km²), both assignments conducted in mainly flat areas and with resolution of less than 10 cm GSD (World Bank Group 2017). Finally, the creation of a digital elevation model for an area of 0.5 km² covering both flat terrain and mountainous and urban areas, using a multirotor UAV with a LiDAR sensor, cost about $14,300. However, in addition to the costs associated directly with the specific data collection assignments, which are mostly a function of the covered area of interest and expected spatial resolution of the outputs, additional, more or less fixed, costs pertaining to predeployment mission planning, flight approvals, and packing and mission rehearsal had to be added, amounting to about $25,500. This cost is consistent with the findings from the World Bank's Technical Guidelines for Small Island Mapping with UAVs (World Bank and Humanitarian OpenStreetMap Team 2019), which suggest that flying the UAV is a very small portion of the time needed; most of the time required is spent in planning, obtaining permission to fly, and postprocessing of the data captured.

UAVs are generally preferable for mapping a small footprint (for example, small pockets of high-flood-risk areas, or a small and remote island community): most mapping drones can cover up to 3 km² in a day or 10 km² in a week of data collection and processing at a cost that is competitive with manned aircraft and satellites. However, satellites are more practical for purposes such as acquiring baseline imagery of large areas at a resolution of 50 cm/pixel with long capture windows (World Bank and Humanitarian OpenStreetMap Team 2019). Especially when ultra-high resolution (less than 10 cm) is not necessary, satellites will likely provide the needed imagery at a much lower price: service providers offer the option of purchasing existing high-resolution archive imagery that can be several days or weeks old for €250–€400 for 25 km² (about $11–$18 per km²), with images delivered in less than four days. Alternatively, new images can be commissioned to be taken once the satellite passes the area in question and weather conditions are suitable, at a cost of $1,500 or $6,500 for minimum areas of 25 km² or 100 km², respectively. For large-scale disasters, the International Charter "Space and Major Disasters" may be activated and free access to satellite-derived information products is then made available to support disaster response.

However, despite the advantages of UAVs in certain contexts, in practice, it is uncommon for a single method to be used exclusively; rather, the various survey methods are used to complement one another. For example, surveys using a satellite or manned aircraft may be conducted every five years across a larger target territory but are complemented by more local-scale UAV survey updates every six months (World Bank and Humanitarian OpenStreetMap Team 2019).

Several drone-based mapping and disaster risk reduction–focused initiatives have been implemented in Tanzania. The census in 2012 estimated the population of Dar es Salaam, Tanzania's largest city, to be 4.36 million; now it is nearly 6 million and rising. The need for better maps is acute because the city is one of the fastest growing in the world, absorbing a thousand people a day (BBC News 2019). And, because many of the areas where newcomers settle do not appear on a detailed map, officially, they do not exist. More than 70 percent of the people in the city live in informal, unplanned settlements with inadequate infrastructure. Heavy rainfalls twice a year result in significant flood risk, with deadly diseases such as cholera a constant threat.

In 2015, a consortium comprising local authorities, COSTECH, two universities, and the Buni Innovation Hub led the production of a detailed map of the city. Aerial imagery at a 5-cm resolution and covering an area of 88 km² was captured over a period of two weeks using fixed-wing UAVs (eBee). The imagery was collected to create digital terrain and elevation models for risk analysis and reduction. With support provided by the Humanitarian OpenStreetMap Team, the data were converted into a complete map of the city, with infrastructure mapped at unprecedented levels of detail. The map is used for urban planning with an emphasis on disaster risk reduction and preparedness for natural calamities, such as floods, as well as health emergencies, such as cholera (Soesilo et al. 2016). Hospitals in Dar es Salaam are also starting to use the maps created by drones to mark where cholera patients are coming from so they can spot where outbreaks are happening.

Supported by the World Bank, the government of Zanzibar has been developing the Open Data for Resilience Initiative (OpenDRI)[10] with the aim of creating evidence-based and innovative solutions to better plan for and mitigate natural disasters. Zanzibar is part of the Southwest Indian Ocean Risk Assessment and

Financing Initiative, which seeks to address the high vulnerability of the Southwest Indian Ocean Island States to disaster losses from catastrophes such as cyclones, floods, earthquakes, and tsunamis. These threats are exacerbated by the effects of climate change, a growing population, and increased economic impacts. Zanzibar's disaster events are mainly related to rainfall, and both severe flooding and droughts have been experienced. The Zanzibar Mapping Initiative created a high-resolution map of the islands of Zanzibar and Pemba, about 2,600 km², using low-cost drones instead of satellite images or manned planes. The maps were created at a very high resolution, 7 cm per pixel, using more than 1,500 drone flights performed over six months. Drone-based approaches to data collection have also helped save money and time in postdisaster contexts, such as after a major landslide in Sierra Leone in 2017 when a rapid damage and loss assessment was conducted to determine reconstruction and recovery needs (see box 3.3).

THE HUMAN IMPACT

Drone-based maps can add value to the planning practices of municipalities and communities and improve the transparency of governance processes, leading to more inclusive and equitable cities. More accurate land maps generated over a shorter period compared with traditional land surveying methods can ensure clarity and protection of property rights and make them more easily transferable, which empowers landowners to become economically

---

BOX 3.3

UAV-based survey of the infrastructure impacts of the Regent-Lumley landslide in Sierra Leone

On August 14, 2017, a massive landslide in the Western Area Rural District of Sierra Leone slipped into the Babadorie River Valley and exacerbated existing flooding, affecting about 6,000 people. The main landslide caused major destruction to infrastructure, including buildings, bridges, schools, and health facilities in the Regent, Malama-Kamayama, Juba-Kaningo, and Lumley areas. The government of Sierra Leone requested the World Bank's support in conducting a comprehensive rapid damage and loss assessment (DaLA), in partnership with the United Nations. The DaLA was carried out from August 24 to September 8, 2017, with the objective of estimating damage and losses and of making preliminary estimates for mobilizing funds and launching immediate recovery.

Along with other data collection methods, the DaLA included an aerial survey of a distance of approximately 6 km to document the extent of the damage along the path of the debris flow (photograph B3.3.1). A fixed-wing, fully autonomous drone, equipped with a camera to collect high-resolution images that can be transformed to 2D orthomosaic and 3D terrain models, was used to do the survey. The assessment was conducted at a resolution of about 8 cm, considered adequate for the purposes of the assessment; the accuracy of the imagery from the drone was increased by supplementing the aerial mapping from each flight with ground control points (GCPs) established before the flight and flagged to be visible from the air. The large area to be covered required more than a single flight, so additional strategies were implemented to avoid challenges such as possible gaps in coverage and damage to the drone during repeated landings; for example, GCP locations and drone launching and

continued

Box 3.3, *continued*

Drone imagery of the Sierra Leone landslide

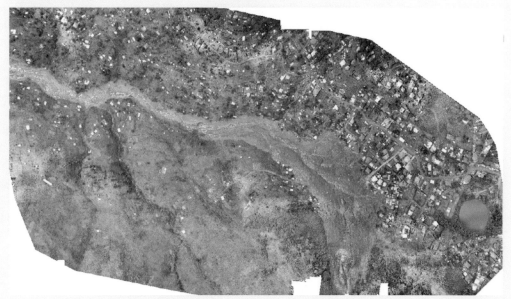

Source: World Bank and GFDRR 2017. © World Bank Arup INTEGEMS. Reproduced with permission from World Bank Arup INTEGEMS Edward Davies Associates LTD; further permission required for reuse.

landing locations were sited in areas that were accessible, relatively flat, and without dense tree cover. The assessment required less than a week at a cost of about $12,000, including the establishment of GCPs and data postprocessing. The main deliverables and benefits produced by the assessment included a detailed remote visual inspection of slip, debris, and damage; a 3D model for design of stabilization and rehabilitation works; and useful data sets for other agencies working on recovery.

stable and provides them with incentives to invest in their land, make it more productive, and consider longer-term planning horizons (Smith and Orçan 2020). Women's property rights have been shown to improve children's health and education, foster inclusive family decision-making, and reduce domestic violence. In the East African context, cost-effective and rapid solutions for gathering data on land tenure also enhance the capacity of government representatives, civil society actors, and individual land owners to effectively deal with pressures on land tenure associated with the extraction of oil and other natural resources in the region.

By storing and demonstrating information about assets, resources, and hazard risks, accurate risk maps enable community members to understand their neighborhoods and to identify community strengths and vulnerabilities to hazards such as floods, landslides, and fires. For example, the detailed ward maps produced by the drone-based flood-mapping initiative in Dar es Salaam in 2015 now enable public health officials to identify specific houses with known cholera patients in case of an outbreak, permitting a much more efficient response through targeted dispatch of medical care

(Soesilo et al. 2016). Availability of lower-cost tools for generating maps can also specifically enhance the resiliency of indigenous communities by helping address issues regarding cultural heritage preservation, economic development, and disaster risk prevention (Li 2020).

Rapid postdisaster needs assessment is key to minimizing the disruptive impacts of flood events. These rapid assessments allow quick estimates to be made of the spatial distribution of demand for water treatment supplies, cholera medicines, and supplies such as temporary latrines to be installed in schools and health centers where displaced families are sheltered (UNICEF 2019). In March 2019, UNICEF used drone imagery to support thousands of families affected by floods in southern Malawi. Many of these people had been forced out of their flooded homes and even more lacked basic supplies such as food, water, and sanitation facilities; thousands of children could not attend school. In emergencies such as this, children are known to suffer the heaviest impact, being at increased risk of malnutrition and disease.

Finally, an additional nonmonetary benefit of using drones for mapping may lie in the associated capacity building and skills development when local pilots are trained and introduced to the technology, eventually creating new business opportunities for them. Such important benefits in the East Africa region have been generated by the Zanzibar Mapping Initiative, among others.

NOTES

1. "A Mobile Application to Secure Land Tenure," https://cadasta.org/a-mobile-application-to-secure-land-tenure/.
2. Also known as an orthomosaic, an orthophoto is an aerial image of an area, composed of multiple photographs stitched together using photogrammetry, which has been scaled and geographically corrected for accuracy. Unlike an uncorrected aerial photograph, an orthophotograph can be used to accurately measure distances.
3. Uganda is among the first countries to fully recognize customary land tenure on the same level with other forms of tenure; customary land ownership is certified using Customary Certificates of Ownership.
4. "OpenDroneMap," https://www.opendronemap.org/odm/.
5. In addition to the UAV and sensor specifications, various factors can affect how large an area can be covered in a single day, including the accessibility of sites for takeoff and landing, maximum flight ceiling, permission to fly beyond line of sight, target overlap between image frames, topography, weather conditions, and other air traffic.
6. "In Tanzania, Drones Are Helping Farmers Stake Their Claim," May 20, 2019, https://blog.werobotics.org/2019/05/20/in-tanzania-drones-are-helping-farmers-stake-their-claim/.
7. In a digital photo (such as an orthophoto) of the ground from air or space, GSD is the distance between pixel centers measured on the ground.
8. Lake Victoria Challenge, www.lakevictoriachallenge.org, accessed April 2019.
9. For example, satellite images in the island context do not have the necessary spatial resolution (pixel size) to show details because the islands are so small relative to pixel size.
10. Data sharing platform: http://zansea-geonode.org; project page: www.zanzibarmapping.com.

REFERENCES

Aditya, T., K. A. Sucaya, F. A. Nugraha, and H. H. S. Lukman. 2020. "Connecting National Land Databases and Para Surveyors through a Mobile Data Collector for Accelerating Land Administration Completeness and Reliability." Paper presented at the Annual World Bank Conference on Land and Poverty, Washington, DC, March 16–20.

Albuquerque, J., B. Herfort, and M. Eckle. 2016. "The Tasks of the Crowd: A Typology of Tasks in Geographic Information Crowdsourcing and a Case Study in Humanitarian Mapping." *Remote Sensing* 8 (10): 859.

Arsanjani, J. J., and E. Vaz. 2015. "An Assessment of a Collaborative Mapping Approach for Exploring Land Use Patterns for Several European Metropolises." *International Journal of Applied Earth Observation and Geoinformation* 35 (March): 329–37.

Bayle, F., N. Kawas, A. Mortarini, C. Rufin, A. Stein, and L. Torres. 2020. "Identification of Climate Change–Related Hazards in Informal Community through the Application of Machine Learning to Satellite Images." Paper presented at the Annual World Bank Conference on Land and Poverty, Washington, DC, March 16–20.

BBC News. 2019. "The Drone Pilot Whose Maps Are Saving Lives in Zanzibar." January 11.

Chadza, C. 2019. "How ATC Interacts with RPAs Operating in Malawi." Malawi Department of Civil Aviation, Lilongwe.

Chingozha, T., D. von Fintel, V. McBride, and K. Govender. 2020. "A Citizen Science Approach to Classifying Urban Informality and Other Urban Land Use Types Using Satellite Imagery." Paper presented at the Annual World Bank Conference on Land and Poverty, Washington, DC, March 16–20.

COSTECH (Tanzania Commission for Science and Technology). 2016. "Spatial Data Delivery Report." Report prepared for the Tanzania Ministry of Lands.

Dijkstra, P., U. Pudasaini, U. S. Panday, E. Unger, and R. Bennett. 2020. "Toward Land Sector Equity and Resilience in Nepal: The Role, Uptake, and Impact of UAVs." Paper presented at the Annual World Bank Conference on Land and Poverty, Washington, DC, March 16–20.

Esch, T., M. Marconcini, F. Bachofer, A. Metz-Marconcini, M. Moenks, J. Zeidler, B. Leutner, E. Anderson, V. Deparday, and C. M. Gavaert. 2020. "Joint Use of Earth Observation, Machine Learning and Local Knowledge to Understand Risk and Increase Resilience in African Cities." Paper presented at the Annual World Bank Conference on Land and Poverty, Washington, DC, March 16–20.

ITF (International Transport Forum). 2018. "(Un)certain Skies? Drones in the World of Tomorrow." International Transport Forum, OECD, Paris.

Koeva, M., M. Muneza, C. Gevaert, M. Gerke, and F. Nex. 2016. "Using UAVs for Map Creation and Updating: A Case Study in Rwanda." *Survey Review* 50 (361): 312–25. doi:10.1080/00396 265.2016.1268756.

Lawani, A., F. E. Mbuya, G. Rambaldi, and H. R. Chaham. 2017. "Unmanned Aerial Systems (UAS): Experts' Synthesis for the High-Level African Panel on Emerging Technologies Meeting." Accra, May 2–3.

Li, Q. 2020. "Integrating Drone, Participatory Mapping and GIS to Enhance Resiliency for Indigenous Communities." Paper presented at the Annual World Bank Conference on Land and Poverty, Washington, DC, March 16–20.

Mbuya, F. E., G. Rambaldi and H. R. Chaham. no date. "Getting Drones Off the Ground in Africa." CTA Policy Brief 11. Technical Centre for Agricultural and Rural Cooperation, Wageningen, the Netherlands.

Noergaard, P., O. S. Kristensen, and K. Sato. 2020. "Efficient Country-Wide Aerial Image Capture as a Foundation for Topographic Mapping, Cadastral Mapping and Land Administration Systems." Paper presented at the Annual World Bank Conference on Land and Poverty, Washington, DC, March 16–20.

Raut, P., R. Shantanu, K. Sushma, and H. Richa. 2020. "Plot Boundary Detection: Smart Agricultural Land Governance Using AI and Satellite Imagery." Paper presented at the Annual World Bank Conference on Land and Poverty, Washington, DC, March 16–20.

Roth, R., and M. Zalba. 2020. "Effective Large Area Mapping of Height Models with Single-Photon LiDAR." Paper presented at the Annual World Bank Conference on Land and Poverty, Washington, DC, March 16–20.

Sanjines Mancilla, M. A., F. Pichel, W. Kambugu, and S. Burke. 2020. "Customary Land Mapping Utilizing Fit for Purpose Approach: Case Study from Buliisa District, Uganda." Paper presented at the Annual World Bank Conference on Land and Poverty, Washington, DC, March 16–20.

Smith, A. H., and B. E. Orçan. 2020. "Generating Geographic Information Systems (GIS) Data and Analytics Acquired Autonomously through Satellites and Unmanned Aircraft to Revolutionise the Development of Modern Land Administration Systems." Paper presented at the Annual World Bank Conference on Land and Poverty, Washington, DC, March 16–20.

Soesilo, D., P. Meier, A. Lessard-Fontaine, J. Du Plessis, and C. Stuhlberger. 2016. "Drones in Humanitarian Action: A Guide to the Use of Airborne Systems in Humanitarian Crises." FSD (Swiss Foundation for Mine Action), Geneva.

UbuntuNet Alliance. 2018. "State University of Zanzibar Shines in Drone Mapping Programme," December 18.

UNICEF. 2017. *Annual Report 2017: Malawi*. Lilongwe, Malawi: UNICEF Malawi.

UNICEF. 2019. "UNICEF Responds to Floods in Southern Malawi, as Number of Affected Families Hits 93,000." Press release, Lilongwe, March 11.

UNICEF Malawi. 2018. "Drones in Malawi." UNICEF Malawi, Lilongwe. https://www.unicef .org/malawi/media/651/file/The%20Drones%20Factsheet%202018.pdf.

UNICEF Malawi. 2019. "Call for Expressions of Interest (EOI): Research, Evaluation, and Market Analysis Services Related to Multi-Purpose Unmanned Aerial Systems (UAS)." UNICEF Malawi, Lilongwe.

USAID (United States Agency for International Development). 2017. *Unmanned Aerial Vehicles Landscape Analysis: Applications in the Development Context*. Global Health Supply Chain Program-Procurement and Supply Management. Washington, DC: Chemonics International Inc. for United States Agency for International Development.

Volkmann, W. 2017. "Small Unmanned Aerial System Mapping versus Conventional Methods: Case Studies on Farmland Surveying." CTA Working Paper 17/07, CTA, Wageningen, the Netherlands.

Volkmann, W. 2020. "Drones in Cadastral Surveying." Paper presented at the Annual World Bank Conference on Land and Poverty, Washington, DC, March 16–20.

Wickless, A., and M. Westers. 2020. "Tackling Undocumented Land Rights Worldwide with Improved Cadastral Survey Productivity." Paper presented at the Annual World Bank Conference on Land and Poverty, Washington, DC, March 16–20.

World Bank and GFDRR (Global Facility for Disaster Reduction and Recovery). 2017. "Sierra Leone Rapid Damage and Loss Assessment of August 14th, 2017 Landslides and Floods in the Western Area." World Bank and Global Facility for Disaster Reduction and Recovery, Washington, DC.

World Bank and GFDRR (Global Facility for Disaster Reduction and Recovery). 2018. "UAV Survey of Bududa Landslide Area." Terms of Reference. World Bank and Global Facility for Disaster Reduction and Recovery, Washington, DC.

World Bank and Humanitarian OpenStreetMap Team. 2019. *Technical Guidelines for Small Island Mapping with UAVs*. Washington, DC: World Bank.

World Bank Group. 2017. "Request for Proposals for Aerial Data Collection in Tonga and Fiji using Unmanned Aerial Vehicles (UAVs)." March 28. World Bank, Washington, DC.

4 Agriculture

THE BIG PICTURE: THE MARKET FOR AGRICULTURAL ASSESSMENTS AND SPECIALIZED SERVICES

Agriculture

The agriculture sector in East Africa is a major provider of incomes and livelihoods and represents close to 50 percent of national gross domestic product (GDP) in some countries. For example, in Kenya the agriculture sector employs more than 40 percent of the total population and more than 70 percent of the rural population. Along with Ethiopia and Tanzania, Kenya has the largest overall area dedicated to maize cultivation and cattle ranching in East Africa (maps 4.1 and 4.2). In Burundi, despite a decline in the sector's contribution to GDP in recent years, agriculture still provides income and jobs to about 85 percent of the population, and the country ranks in the 98th percentile of the global distribution of agriculture's share of GDP (World Bank 2018). Sub-Saharan Africa overall needs to double (and perhaps even triple) current levels of agricultural productivity to meet demand and stave off food and nutrition insecurity (CTA 2019). Across Sub-Saharan Africa, efforts to increase farmers' opportunities to access credit are growing. The provision of detailed and up-to-date spatially defined data on farm location and size and the health and biomass of standing crops can help improve farmers' creditworthiness. In countries such as Malawi, recurring droughts have ravaged the agriculture sector, threatening the livelihoods of smallholder farmers (producing crops or livestock on two or fewer hectares of land), who constitute 80 percent of the country's population (Oakland University 2018). Drones can potentially increase the security of livelihoods by enabling better prediction of seasonal and environmental patterns and recommendations for solutions to low crop yields using data gathered via multispectral aerial images.

Artificial insemination of cattle

The dairy sector in East Africa is large and growing and is a critical source of income and nutrition for millions of low-income people. The dairy cow population in the extended East Africa region is more than 13 million, ranging from

MAP 4.1
East Africa: Area of maize harvested

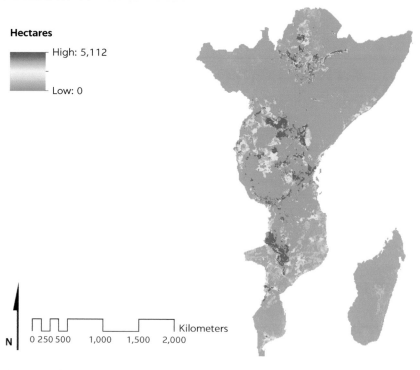

Source: Data from IFPRI 2017.

MAP 4.2
East Africa: Cattle density

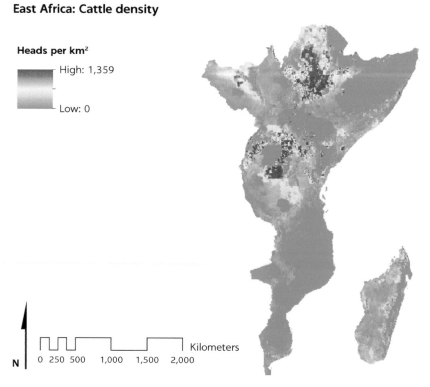

Source: Data from IFPRI 2017.

20,000 in Malawi to 3.8 million in Kenya and 4.5 million in Ethiopia. Milk production per cow is much higher in Kenya, Madagascar, Malawi, and Tanzania (more than 1,000 liters per year), compared with Burundi, Ethiopia, Rwanda, or Uganda (less than 650 liters per year). Per capita annual milk consumption varies from just 3.7 liters in Malawi and 4.5 liters in Burundi to 21.5 liters in Uganda and 76.7 liters in Kenya. These differences in milk consumption represent differences in micronutrient availability with far-reaching consequences, especially affecting women and children. They reflect differences in milk production efficiency and therefore milk availability. Even the best performers in the region are still far below the possible performance of dairy systems in Africa (Kemp and Harvey, n.d.). A limiting factor for dairy production in East Africa is genetics and fertility. A cow that is not successfully inseminated at the first opportunity leads to lost production and a cost to the farmer of several dollars per day for at least one month. The opportunity to improve the genetic quality of the national herd depends on getting the right semen, in the right condition, to the right cow during the critical 24 hours of heat. In 2015 the total value of Kenya's cattle artificial insemination business, used in 18 percent of total dairy cattle breeding, exceeded $11 million, with the potential to generate more than $37.6 million at 60 percent of total dairy cattle breeding. Artificial insemination is used predominantly in dairy cattle, whereas less than 1 percent of the beef herd is bred using artificial insemination. The use of artificial insemination is projected to reach up to 2.3 million inseminations per year by 2023 compared with 650,000 in 2015 (Makoni, Hamudikuwanda, and Chatikobo 2015). The Rwanda Livestock Master Plan considers artificial insemination and synchronization, combined with improved feed and health interventions, to be one of the priorities for increasing national cow milk production (Shapiro et al. 2017).

In addition to the logistical challenges in the cattle artificial insemination sector, another factor affecting dairy sector productivity in East Africa is the lack of availability and affordability of vaccines, drugs, and diagnostics when and where needed. Both challenges are supply chain last-mile gaps that have large impacts and that unmanned aerial vehicles (UAVs) could help address.

PRESENT COSTS AND METHODS FOR AGRICULTURAL ASSESSMENTS AND SERVICES

Agricultural techniques

Precision agriculture techniques have until now been accessible only to large-scale farmers for whom the economies of scale warranted the investment in what used to be expensive hardware and services. Precision agriculture is a way to apply interventions in the right place at the right time (Gebbers and Adamchuk 2010). Most conventional satellite technologies are developed with mechanized conventional agriculture in mind—the large rectangular fields and single crops that are common in developed countries—whereas farming practices are very different in Sub-Saharan Africa: plots are considerably smaller, farmers grow multiple crops with similar plant cycles, and intercropping is a common practice (Hall and Archila Bustos 2016).

Many low- and middle-income countries lack the infrastructure and resources to conduct frequent and extensive agricultural field surveys to obtain quantitative and up-to-date information on the types of crops being cultivated, the acreage under cultivation, and crop yields—data that are needed by stakeholders at all scales (local, regional, and national) to make decisions about agricultural production, development priorities, and policies (Temple et al. 2019). These information gaps apply to stakeholders in East Africa as well. Moreover, many farmers in the region continue to lack robust information on more nuanced issues such as their exposure to drought risks, which partly reflects an absence of data. For example, meteorological data are currently collected from expensive and unreliable metering stations and provide little spatially granular understanding of the relationship between crop production and rainfall variations.

However, innovative solutions to this problem that leverage free satellite data are starting to become available. For example, as part of the UK Space Agency's International Partnership Programme,[1] a recent project in Kenya analyzed free Earth observation data from the European Space Agency's Sentinel-2 satellite for drought resilience monitoring, in combination with space-derived weather and temperature data from the Moderate Resolution Imaging Spectroradiometer. These inputs are then processed with an innovative algorithm to generate a Vegetation Health Index every 10 days throughout the growing season, thus providing a very detailed (20-meter ground resolution) assessment of crop development and potential yields. Analysis of the project suggests that the satellite data–based approach is a cost-effective solution (Deane and Vescovi 2020). However, acquiring useful satellite imagery may be challenging in individual parts of the region; for example, clouds can block images acquired using light in the visible and infrared ranges. Temple et al. (2019) find that, in Rwanda, where cloud cover during the two main growing seasons is significant, only 10 percent of the Sentinel-2 35-MRT[2] image tiles (covering one-third of Rwanda) available for 2018 were cloud free.

Artificial insemination services

The livestock artificial insemination services sector is entirely or mostly public in most countries in the region (Burundi, Ethiopia, Rwanda, Tanzania), although some private sector initiatives are emerging. Artificial insemination of livestock in Tanzania is supervised by the National Artificial Insemination Center, supported by funding from the Bill & Melinda Gates Foundation. In 2015, this five-year initiative with a budget of $18.1 million set a goal of supporting 800 private and public artificial insemination service providers to train at least 225,000 smallholders on improved dairy cattle management and to deliver approximately 1.8 million artificial insemination and other dairy production–related inputs and services in Ethiopia and Tanzania (Land O'Lakes International Development Fund 2015).

In Rwanda, the government subsidizes the use of artificial insemination, and farmers seem to be aware of the program. Roughly 3.75 percent of farmers use exclusively artificial insemination, and 21.35 percent use both artificial insemination and natural mating in cattle breeding (Mazimpaka et al. 2017). Access to the artificial insemination centers seems to be one of the main factors delaying the success of the initiative. According to a study conducted in Nyagatare, Rwanda, 79.2 percent of respondents residing within 5 kilometers (km) of the artificial insemination centers adopted artificial insemination, whereas 92.5 percent of respondents residing more than 15 km from the centers preferred the natural service (Mushonga et al. 2017). In Uganda, artificial insemination is supervised by the National Animal Genetics Resources Centre and Data Bank. Countrywide, more than 1,840 people are trained as artificial insemination technicians. More than 80 percent of the districts in the country are supplied with artificial insemination equipment under National Agricultural Advisory Services (Nasasira 2017).

However, despite the favorable institutional environment, the use of artificial insemination at scale across East Africa requires that various supply chain challenges be overcome. Cattle artificial insemination requires that semen in liquid nitrogen storage reach the cow within a critical time window, meaning there is a need for constant availability of liquid nitrogen and semen in remote areas, a network of expensive and fragile liquid nitrogen plants, and a network of artificial insemination technicians whose business is dependent on maintaining stocks of frozen semen. These technicians face big risks, high transaction costs, and small profits. As the semen and liquid nitrogen system has collapsed, technicians have disappeared. Farmers cannot get the artificial insemination they need, when they need it, and in the required condition. As a result, artificial insemination in the East Africa region is significantly underused. Farmers rely on natural service, which is very inefficient and expensive and cannot drive genetic improvement.

INTEGRATING DRONES INTO EXISTING AGRICULTURAL DATA COLLECTION AND SERVICES SYSTEMS

In what types of applications and contexts can UAVs add value?

Agriculture

At the annual African Green Revolution Forum,[3] global and African leaders come together to develop actionable plans that will move African agriculture forward, including by supporting African smallholder farmers to more quickly adapt to challenges such as recurrent droughts and emerging pests. In 2019, the forum took place in Ghana under the theme "Grow digital: Leveraging digital transformation to drive sustainable food systems in Africa." UAVs were among the practical applications emerging from the digital revolution that were explored, along with big data, blockchain, machine learning, robotics, and sensors.

Agricultural aircraft have been in use since the 1920s, and the use of satellites to assess crop health from the sky has increased in recent decades. Although research has demonstrated the ability of satellite imagery combined with artificial intelligence methods to generate useful data in contexts in which smallholder systems dominate (Burke and Lobell 2017), UAVs likely represent the next stage in the progression from the "macro" to the "micro," or from large-scale to small-scale farms (Greenwood 2016). UAV applications in agricultural vegetation monitoring date back nearly two decades, with examples including large fixed-wing UAVs tested for cash crops such as coffee and vineyards in the early 2000s (Berni et al. 2008). The continuing reductions in the cost, size, and complexity of UAVs and the associated sensors mean that African farmers could leapfrog from more traditional methods of agriculture to so-called precision agriculture (Mbonyinshuti 2016). In particular, UAVs are expected to provide significant help to farmers in developing countries, who historically have found it harder to access aerial imagery, either from manned aircraft or from satellites (Greenwood 2016). The general trend across many of the precision agriculture advisory solutions in Africa, particularly as the costs of underlying technologies decrease, is toward fully integrated precision advisory platforms that combine not only drone data but also in-depth farmer profiles, transaction data, weather data, satellite data, and field and machinery sensor data (CTA 2019). Examples of such platforms include Microsoft's FarmBeats in Kenya and Tata Consultancy Services' InteGra precision agriculture advisory platform in South Africa, in addition to a number of start-ups.

Some of the most exciting opportunities for UAV technology overall lie in improving the management of crops, livestock, fisheries, forests, and other natural resources. In addition to precision agriculture applications, UAVs offer a range of promising ways to enhance the management of agricultural assets more broadly: they can improve the monitoring of livestock, fences, trees, and wildlife, and help farmers document their farms to improve their creditworthiness. They can also allow insurers to carry out rapid crop insurance assessments, even after major disasters. UAVs can give farmers a birds-eye view of their crops, permitting them to detect subtle changes that cannot be identified by "crop scouts" at the ground level. The growing range of UAVs on the market are capable of collecting high-resolution data through image, infrared, thermal, and chemical sensors. UAVs equipped with specialized sensors can collect multispectral images that are used to generate data on crop quality, such as the Normalized Difference Vegetation Index.[4] The data can also be used to speed up the process of conducting crop inventories and yield estimates. Moreover, multispectral imaging enables drones to pick up on crop conditions that are not always clear to the human eye, such as water stress, fertilizer needs, pest infestations, and diseases. The Food and Agriculture Organization has begun to investigate how drones could be used to detect and eliminate agricultural pests such as locusts, given the enormous costs and devastating impacts associated with local plagues that are not caught in time; for example, it took more than $500 million and two years to control the 2003 and 2005 locust crises in northern Africa (Cressman 2016).

Even equipped with consumer-grade cameras, UAVs flying at an altitude of 100 meters can produce images with a spatial resolution of 3-4 centimeters, which is much higher than the spatial resolution available from most satellite platforms. With the spatial resolution attained with UAV-based remote sensing, crop discrimination is feasible even with conventional RGB (red, green, blue) cameras. Unlike most satellites, UAVs also facilitate the capture of high-resolution imagery at frequent time intervals, which may be important in crop monitoring; unlike manned aircraft, UAVs can do it at a reasonable cost and level of operational complexity (Berni et al. 2008). Cattle ranchers can use UAVs to determine where their livestock are in nearly real time, and some have found UAVs useful for conducting regular surveys of fencing (Greenwood 2016).

According to the drone companies operating in East Africa interviewed for this report, a large share of the UAV-based initiatives to date, in both crop monitoring and precision agriculture (including crop spraying), have been funded by external donors rather than through national government program funds. Similarly, despite the potential at smaller scales, large-scale farmers remain the main clients for agricultural UAV applications in the region, partly because the benefits at scale are more tangible and measurable. As also pointed out by the interviewed drone operators, increased scaling of the UAV agricultural use cases in the region would benefit not only the farmers but also the local drone companies, given that the agricultural applications provide an opportunity for drone companies to relatively easily build up flight hours because of the low associated risk of rural inspections compared with, for example, urban flood risk assessments.

In Malawi, high-precision aerial photography captured by drones and weather station data have been used to analyze and predict crop yields as part of Feed the Future, the US government's global hunger and food security initiative funded by the United States Agency for International Development. The captured imagery has allowed researchers to quantify how much water and chlorophyll are in the plants and to conduct three-dimensional measurements of plants in different parts of the field, enabling the development of more appropriate solutions to low crop yields. The researchers are also working to develop models that can better predict seasonal and environmental patterns that have been disrupted by climate change (Oakland University 2018).

In Tanzania, Agrinfo partnered with WeRobotics and Tanzania Flying Labs in 2018 to gather field data in rural areas of the country for the NASA Harvest consortium's Pre-Harvest Loss for Smallholder Farmers initiative in collaboration with the International Food Policy Research Institute and the University of Maryland (WeRobotics 2018). As part of the Spurring a Transformation for Agriculture through Remote Sensing project led by the University of Twente in the Netherlands and funded by the Bill & Melinda Gates Foundation, a team from the University of Maryland used UAV technology—two fixed-wing drones— to support the collection of national agricultural statistics in Tanzania by mapping maize-based agricultural systems (Zurita Milla 2016). The results from the UAV flights were scaled up to the national level using satellite data and crowdsourced information from the ground. The resulting cropland map was then shared with the Tanzanian Ministry of Agriculture to support its food security policy making.

Cattle insemination

UAV use in cattle artificial insemination has been demonstrated in a collaboration between the Rwandan government and Zipline in Rwanda. Unit costs of the model can be reduced by synchronizing groups of recipient cows with drone-delivered hormones, collaborating with human health deliveries to bundle cargos (in the test phase in Rwanda, Zipline's network of clinics was used as drop sites for semen), and, as dairy profitability increases, bundling additional veterinary products (Kemp and Harvey, n.d.).

Challenges

Overall, although UAVs are a promising technology across several agricultural applications, several obstacles must still be overcome before they become a standard part of the farmer's toolkit. CTA (2019) estimates that across the more than 30 smallholder-focused drone start-ups in Africa, only a few hundred thousand hectares of land have been scanned and, likely, only tens of thousands of African farmers have had their fields analyzed via drone flyovers in the past few years. In addition to regulatory uncertainty, another barrier to UAV adoption in agriculture in many parts of Sub-Saharan Africa currently lies in lack of awareness and education: although drones are relatively easy to use, farmers will still need training in their local language and technical support, as well as up-to-date information on the UAV technologies' legal status in their country (Greenwood 2016). Variable availability of support infrastructure presents another challenge in parts of the region, given that UAVs' reliance on communications from a ground operator for control makes them vulnerable to signal loss from interference, flying out of range, and hacking (African Union and NEPAD 2018).

Potential time and cost savings

Drones have already proven useful to agricultural planners, greatly reducing the time and cost required to conduct accurate crop surveys (Greenwood 2016). When used in precision spraying of crops, drones provide higher flight speed and greater spraying flow, reaching operational efficiency of up to 14 hectares per hour, equivalent to the work of 100 farmland laborers. The detection of crop variability using drone-based imagery and drone-based crop spraying can be accomplished as much as 40 times faster than when using conventional methods according to Charis,[5] a drone company that provides services to private sector and government agencies in Rwanda and also has contracts in neighboring countries and Côte d'Ivoire. Charis, in partnership with the University of Rwanda, has been conducting detailed cost-benefit analyses of drone-based precision agriculture initiatives since 2020; this research will provide insights into the exact value added this service provides to the farmers and their willingness to pay.

Despite the up-front speed advantages, the data collected by drones equipped with sensors does not automatically equal "information," which is what agricultural producers ultimately require; hence, whatever data are captured by drones need to be converted into actionable information for farmers, with associated additional time requirements (African Union and NEPAD 2018). For example, in agricultural crop inspection in Kenya, the drone operators interviewed for the current study reported large time savings compared with traditional inspection methods: the agricultural drone is typically capable of covering as much as 1,200 acres per day. However, the operators also pointed out that image acquisition must be followed by significant time dedicated to image processing; as a rule of thumb, 1,000 acres worth of images captured requires 15 hours of time for image processing (typically run offline on a powerful computer). Yet additional time requirements, also borne by the drone operators themselves, are associated with educating local farmers on the specifics of drone-based inspection and its advantages.

Work underway at the International Maize and Wheat Improvement Center (CIMMYT) in Zimbabwe is seeking to ensure that the widespread hunger in the country caused by the 2015–16 drought is not repeated by breeding a heat- and drought-tolerant maize variety that can still grow in extreme temperatures. CIMMYT maize breeders use climate models from the global Consultative Group on International Agricultural Research Program on Climate Change, Agriculture and Food Security to inform breeding decisions such as by identifying the genetic merit of each individual plant so that the best ones can be selected for breeding. In its search for cost-effective ways to assess a larger number of maize plants and to collect more accurate data related to key plant characteristics, CIMMYT is turning to UAVs. Multispectral and thermal images taken by cameras on UAVs by CIMMYT researchers in other countries are already helping monitor the resistance of maize to tar spot complex and other foliar diseases. The research shows that this approach, known as "phenotyping," can speed up and improve the effectiveness of disease assessment in experimental maize plots (Cowan 2019). Under the Stress Tolerant Maize for Africa[6] project in Zimbabwe, CIMMYT researchers are working on implementing the use of drone-based sensing, among other breeding innovations, to reduce the time and cost of phenotyping so that the development of new varieties costs less. The use of UAVs cuts time and cost of data collection by 25–75 percent compared with conventional methods because it enables data on several traits, for example, canopy senescence and plant count, to be collected simultaneously (Chikulo 2019).

In Nigeria, drone technology has accelerated the planning, design, and construction of rice irrigation systems. Teams planning a 3,000-hectare irrigated rice farm near New Busa, some 700 km from the country's capital, Abuja, used drone imagery to make decisions on the layout of both rice paddies and irrigation and drainage systems. Thanks to the drone imagery, they were able to quickly determine that their original design was poorly suited to the available terrain. Although a manned aircraft could have done the job, it also would have been prohibitively expensive. With the UAV, nearly 300 hectares of land could be examined during the 55-minute flight, and nearly 1,000 hectares could be covered in a single day; in comparison, it would have taken a professional surveyor working on foot about 20 days to cover the same area (Le 2016). A similar UAV application in Ghana is illustrated in photograph 4.1.

The interviewed drone operators suggested that the realistic all-inclusive cost per acre of crops inspected by drones is currently $3 (including aircraft depreciation, crew costs, maintenance, and fuel), in addition to a surcharge of $500 per 1,000 acres. The frequently cited $0.50 per linear km flown across drone use cases was highlighted as not including research and development or drone amortization costs. When compared with satellite image–based crop inspection (about $0.40 per acre), drone-based costs are higher; however, the typical resolution of images obtained by UAVs is significantly higher as well,

PHOTOGRAPH 4.1

On-field drone demonstrations at the Kpong irrigation scheme in Ghana, sponsored by CTA

Source: Africa Green Revolution Forum 2019, © CTA ACP-EU, CC BY-SA 2.0, https://www.flickr.com /photos/cta-eu/48741981093/. Used with permission; further permission required for reuse.
Note: CTA = Technical Centre for Agricultural and Rural Cooperation.

which is not achievable with satellite data for various reasons, such as cloud cover. Unlike drone-based imagery, the more affordable satellite images typically lack the level of resolution that would permit distinguishing between plant types.

The drone operators also pointed out that in the rural East Africa context, physically getting to the farms to be inspected is often the most expensive part of the work, with 200 km in travel distance being the common upper limit for whether the assignment is accepted. Individual operators use formal proprietary cost models to determine whether an assignment can generate positive earnings given the various costs associated with operating in specific contexts. Box 4.1 provides an example of the complicated process of defining a viable business model for drone-based data provision in the small-scale farming sector in Mozambique.

BOX 4.1

Defining a business model for drone-based data provision to smallholder farmers in Mozambique

The ThirdEye: Flying Sensors to Support Farmers' Decision Making project in Mozambique's Xai-Xai and Chókwè districts began in 2014 with funding from the United States Agency for International Development, the Swedish International Development Cooperation Agency, and the Ministry of Foreign Affairs of the Netherlands. Extension services increasingly delivered by private advisory service providers worldwide rarely serve the rural poor; in contrast, the ThirdEye project aims to be a self-sustaining company delivering its Flying Sensor services to smallholder farmers in particular. The project uses recreational unmanned aerial vehicles (UAVs) equipped with near-infrared sensors and tailored software to locally capture and analyze data.

The original business model focused on supporting smallholder farmers in exchange for payment for the services provided, with the total costs for an end user (assuming the average farm size in Mozambique of 0.5 hectare) estimated at $0.95–$1.00 per farmer per month. However, subsequent surveys with target groups revealed that the willingness of farmers to pay for the services is very low, at up to $0.40 per month. The seasonal nature of agricultural income and payment collection presented an added complexity. In other words, a business plan solely focused on sales of services to smallholder farmers is not financially feasible for ThirdEye. Research by van den Akker (2016) examined several alternative business models that

would enable ThirdEye to move from a donor-funded project to a profitable company while maintaining its social focus on smallholder farmers. These models included the following, among others:

- Including large-scale commercial farmers in the customer segment (to account for up to 40 percent of all customers), given that they are willing and ready to pay for the services delivered by ThirdEye.
- Receiving funding for services to farmers from organizations or companies further along the value chain: irrigation schemes, agribusiness firms, and nongovernmental organizations are examples of value chain actors who benefit from increased agricultural output and productivity of farmers, including smallholders.
- Having agro-input providers cover part of the costs: organizations producing and distributing agricultural inputs (fertilizers, seeds, pesticides, agricultural equipment, and so on) may use the Normalized Difference Vegetation Index information generated by ThirdEye to show the need for their products and help gain the trust of the smallholder farmers.

Overall, the study concluded that, to be financially sustainable, the project's customer focus should shift toward a system in which smallholder farmers would be the main end user but not the main paying customer.

The unfulfilled demand for cattle artificial insemination in East Africa could potentially be a viable use case for UAVs, given the high-value cargo with a short shelf life, the need to deliver the cargo within a tight time frame, large short- and long-term benefits derived from meeting this need, and the existing data network of potential clients (such as in Ethiopia and Tanzania). UAVs with a one-way range of about 180 km would provide an alternative, more robust and responsive hub-and-spoke distribution model that removes the need for hundreds of remote, expensive, fragile liquid nitrogen plants. Specifically, the drone-based supply chain would consist of the following:

- Distribution of livestock semen in dry ice in packages with a frozen life span of up to 24 hours; dry ice can be made on demand, on site from liquid carbon dioxide, which is readily available and keeps indefinitely at no cost; and
- The artificial insemination technician receiving the package close to the client and conducting the insemination the same day, thus removing the need for and the associated costs of stocking semen or liquid nitrogen.

However, although there are no apparent technical limitations to using UAVs in cattle artificial insemination, the business model is hard to demonstrate because the current system relies on a complex mix of public sector actors and very small-scale private sector artificial insemination technology firms and farmers. As with all use cases, the economic feasibility of the approach is sensitive to the local density of dairy cattle within drone range. Moreover, real-time links must be established between the farmers, the artificial insemination technicians, and the supply chain.

THE HUMAN AND ENVIRONMENTAL IMPACTS OF UAVs

The important role that UAVs can play in enhancing global food security while reducing the stress on the environment is widely recognized (Fleming 2016; Weilbach 2016). In Africa, digitalization for agriculture, of which UAVs are one component, can be a game changer in supporting and accelerating agricultural transformation across the continent (CTA 2019). The adoption of farming practices that will increase yields with lower inputs while optimizing profit provides the basis for well-being and poverty reduction (African Union and NEPAD 2018). The application of drone technology in agriculture can help generate additional revenues and save resources for local farmers; for example, the detection of crop variability and quality can increase farmer profits by helping increase the efficiency of irrigation and fertilizer application and avoiding crop loss. Estimates suggest that a half-ton increase in staple yields per hectare alone could generate 13–20 percent higher GDP per capita in many countries (McArthur and McCord 2014). Improved access to crop health information can also help enable the rural micro-insurance industry to reduce the financial impact of drought events, improving the stability of farmers' livelihoods. Improved agricultural data can economically empower Africa's women, who constitute 40–50 percent of the continent's smallholder producers. For society at large, agricultural transformation, enabled by drones and other digital technologies and data, will likely result in lower prices, and improved market links will result in greater access to nutritious food (CTA 2019). Finally, improved agricultural information could be used to detect pockets of rapidly increasing food insecurity, enabling more effective policy responses to be developed that can alleviate political instability in areas that rely heavily on agricultural production (Temple et al. 2019).

Examples of tangible benefits provided by the use of UAVs in agriculture for local producers are beginning to emerge in several countries in the region. As of 2017, the ThirdEye: Flying Sensors to Support Farmers' Decision Making project in Mozambique had trained 14 local extension workers to use UAVs to generate information for 2,800 smallholder farmers, mostly women, covering 1,800 hectares of land, improving the farmers' decision-making regarding when to plant, fertilize, and irrigate. The total number of beneficiaries of the initiative is an estimated 14,000, and farmers have recorded a 41 percent increase in crop production (de Klerk et al. 2017).

The use of drones for precision agriculture also potentially provides large benefits by improving the environmental sustainability of agricultural operations. Precision agriculture is crucial to minimizing environmental damage from agriculture, for example, by reducing nitrate leaking, improving water-use efficiency, and increasing fuel efficiency (Zarco-Tejada, Hubbard, and Loudjani 2014). With regard to water-use efficiency, precision agriculture reduces wastage compared with uniform spraying of water or other irrigation systems (Hendriks 2011). In Morocco, UAV services are used to help ensure that fertilization efficiency and desired quality are met. For cereals, doing more with less is possible thanks to accurate fertilization advice that helps determine the exact nitrogen needs of each square meter of a parcel. In a recent project in Morocco, thanks to variable rate fertilizer application, 2.5 fewer tons of ammonium nitrate were used over a total area of 78 hectares while yield was maintained (Lawani et al. 2017). Similarly, the ThirdEye project in Mozambique resulted in a 55 percent improvement in water productivity (de Klerk et al. 2017), the standard international measure of water savings (for example, Perry 2011). In total, the ThirdEye service led to a reduction in water use of more than 3.7 million cubic meters.

NOTES

1. "International Partnership Programme," www.gov.uk/government/collections/internatio nal-partnership-programme.
2. MRT stands for "MODIS Reproject Tool," a software program.
3. "African Green Revolution Forum (AGRF) 2019," https://www.cimmyt.org/events/african -green-revolution-forum-agrf-2019/.
4. The Normalized Difference Vegetation Index is a simple graphical indicator that can be used to analyze remote sensing measurements assessing whether the target being observed contains live green vegetation.
5. Charis UAS, https://charisuas.com/latest-news/.
6. "Stress Tolerant Maize for Africa," https://stma.cimmyt.org/.

REFERENCES

African Union and NEPAD (New Partnership for Africa's Development). 2018. *Drones on the Horizon: Transforming Africa's Agriculture.* Gauteng, South Africa: NEPAD.

Berni, J. A. J., P. J. Zarco-Tejada, L. Suárez, V. González-Dugo, and E. Fereres. 2008. "Remote Sensing of Vegetation from UAV Platforms Using Lightweight Multispectral and Thermal Imaging Sensors." Unpublished. https://www.isprs.org/PROCEEDINGS /XXXVIII/1_4_7-W5/paper/Jimenez_Berni-155.pdf.

Burke, M., and D. B. Lobell. 2017. "Satellite-Based Assessment of Yield Variation and Its Determinants in Smallholder African Systems." *Proceedings of the National Academy of*

Sciences of the United States of America 114 (9): 2189–94. https://doi.org/10.1073/pnas.1616919114.

Chikulo, S. 2019. "Digital Imaging Tools Make Maize Breeding Much More Efficient." International Maize and Wheat Improvement Center (CIMMYT), March 11, 2019. https://www.cimmyt.org/news/digital-imaging-tools-make-maize-breeding-much-more-efficient/.

Cowan, C. 2019. "Bird's-Eye View." International Maize and Wheat Improvement Center (CIMMYT), June 20, 2019. https://www.cimmyt.org/news/birds-eye-view/.

Cressman, K. 2016. "Preventing the Spread of Desert Locust Swarms." *ICT Update* 82 (April): 8–9.

CTA (Technical Centre for Agricultural and Rural Cooperation). 2019. *The Digitalisation of African Agriculture Report 2018–2019*. Wageningen, the Netherlands: CTA.

Deane, G., and F. Vescovi. 2020. "Monitoring Crop Health from Space to Improve Drought Resilience for Farmers in Kenya." Paper presented at the Annual World Bank Conference on Land and Poverty, Washington, DC, March 16–20.

de Klerk, M., P. Droogers, G. Simons, and J. van Til. 2017. "Change in Water Productivity as a Result of ThirdEye Services in Mozambique." FutureWater Report 166, FutureWater, Wageningen, the Netherlands.

Fleming, N. 2016. "How to Feed 9.7 Billion? Startups Take on the Global Food Problem." *The Guardian*, June 1, 2016. http://www.theguardian.com/world/2016/jun/01/how-to-feed-9-billion-people-startups-global-food-problem-agriculture-venture-capital.

Gebbers, R., and V. Adamchuk. 2010. "Precision Agriculture and Food Security." *Science* 327 (5967): 828–31.

Greenwood, F. 2016. "Drones on the Horizon: New Frontier in Agricultural Innovation." *ICT Update* 82 (April): 2–4.

Hall, O., and M. F. Archila Bustos. 2016. "The Challenge of Comparing Crop Imagery Over Space and Time." *ICT Update* 82 (April): 14–15.

Hendriks, J. 2011. "An Analysis of Precision Agriculture in the South African Summer Grain Producing Areas." North-West University, Potchefstroom, South Africa.

IFPRI (International Food Policy Research Institute). 2017. CELL5M: A Multidisciplinary Geospatial Database for Africa South of the Sahara. HarvestChoice, International Food Policy Research Institute (IFPRI); University of Minnesota. https://doi.org/10.7910/DVN/G4TBLF.

Kemp, S., and D. Harvey. No date. "The Niche for Drones in the Dairy Sector Supply Chain." Unpublished presentation.

Land O'Lakes International Development Fund. 2015. "New Program to Transform AI Service Delivery and Dairy Production in East Africa." Cision, PR Newswire, November 16, 2015. https://www.prnewswire.com/news-releases/new-program-to-transform-ai-service-delivery-and-dairy-production-in-east-africa-300179561.html.

Lawani, A., F. E. Mbuya, G. Rambaldi, and H. R. Chaham. 2017. "Unmanned Aerial Systems (UAS): Experts' Synthesis for the High-Level African Panel on Emerging Technologies Meeting." Accra, May 2–3.

Le. Q. 2016. "A Bird's Eye View on Africa's Rice Irrigation Systems." *ICT Update* 82 (April): 6–7.

Makoni, N., H. Hamudikuwanda, and P. Chatikobo. 2015. "Market Study on Artificial Insemination and Vaccine Production Value Chains In Kenya." Unpublished. https://agriprofocus.com/upload/Market_Study_on_AI_and_Vaccine_Production_Value_Chains_in_Kenya1429877294.pdf.

Mazimpaka, E., F. Mbuza, T. Michael, E. Gatari, E. Bukenya, and O.-A. James. 2017. "Current Status of Cattle Production System in Nyagatare District-Rwanda." *Tropical Animal Health and Production* 49 (8): 1645–56. https://doi-org.ezproxy.cul.columbia.edu/10.1007/s11250-017-1372-y.

Mbonyinshuti, J. D. 2016. "Drone to Monitor Crops in Northern Province." The New Times. https://goo.gl/5d0fka.

McArthur, J., and G. C. McCord. 2014. "Fertilising Growth: Agricultural Inputs and Their Effects in Economic Development." Brookings Institution, Washington, DC.

Mushonga, B., J. Dusabe, E. Kandiwa, E. Bhebhe, G. Habarugira, and A. Samkange. 2017. "Artificial Insemination in Nyagatare District: Level of Adoption and the Factors Determining Its Adoption." *Alexandria Journal of Veterinary Sciences* 55 (1): 1–7. doi:10.5455/ajvs.273226.

Nasasira, R. D. 2017. "Artificial Insemination to Boost Local Animal Production." Daily Monitor, July 3, 2017. https://www.monitor.co.ug/Magazines/Farming/Artificial-insemination-to -boost-local-animal-production/689860-3998334-2vy2lvz/index.html .

Oakland University. 2018. "Anthropology Professor Deploys Drone to Combat Hunger in Africa." March 8. https://www.oakland.edu/socan/news/2018/anthropology -professor-deploys-drone-to-combat-hunger-in-africa.

Perry, C. 2011. "Accounting for Water Use: Terminology and Implications for Saving Water and Increasing Production." *Agricultural Water Management* 98 (12): 1840–46. http://doi .org/10.1016/j.agwat.2010.10.002.

Shapiro, B., G. Gebru, S. Desta, and K. Nigussie. 2017. "Rwanda Livestock Master Plan." International Livestock Research Institute, Nairobi. http://extwprlegs1.fao.org/docs/pdf /rwa172923.pdf.

Temple, D. S., J. S. Polly, M. Hegarty-Craver, J. I. Rineer, D. Lapidus, K. Austin, K. P. Woodward, and R. H. Beach. 2019. "The View from Above: Satellites Inform Decision-Making for Food Security." RTI Press Publication No. RB-0021-1908, Research Triangle Park, NC. https://doi .org/10.3768/rtipress.2019.rb.0021.1908.

van den Akker, J. 2016. "Identifying and Designing Business Models for Innovative Flying Sensor Services in Mozambique." A research project Submitted to Van Hall Larenstein University of Applied Sciences in partial fulfilment of the requirements for the bachelor degree in Sustainable Value Chain.

Weilbach, F. 2016. "Africa: Tech Innovation Will Catalyse Productivity and Growth in Africa Says PWC Report." May 31. http://allafrica.com/stories/201605310931.html.

WeRobotics. 2018. "Tanzania Drone Pilots Team up with IFPRI and Local Smallholder Farms." Blog. June 14, 2018. https://blog.werobotics.org/2018/04/10/tanzania-drone-pilots -team-up-with-ifpri-and-local-smallholder-farms/.

World Bank. 2018. "Republic of Burundi: Addressing Fragility and Demographic Challenges to Reduce Poverty and Boost Sustainable Growth." Systematic Country Diagnostic. Report 122549-BI, World Bank, Washington, DC.

Zarco-Tejada, P. J., N. Hubbard, and P. Loudjani. 2014. "Precision Agriculture: An Opportunity for EU Farmers-Potential Support with the CAP 2014-2020." Joint Research Centre (JRC) of the European Commission, Ispra, Italy.

Zurita Milla, R. 2016. "Transforming Smallholder Farming through Remote Sensing." *ICT Update* 82 (April): 18–19.

5 Infrastructure Inspection

THE BIG PICTURE: REGIONAL DEMAND FOR INFRASTRUCTURE INSPECTION

The need for transport infrastructure inspection globally and in East Africa specifically is vast, especially considering the backlog. Road agencies or authorities are responsible for verifying the geometry and inspecting the condition of the primary and secondary road networks and all related structures. The rough rule of thumb is to cover a proportion of the network every year: in most countries, anywhere between 20 percent and 40 percent of well-managed road networks are inspected in a given year. Some types of inspections are performed on an even more regular basis; in the Democratic Republic of Congo, for instance, road agencies are required to provide an update on the status of a road network of about 58,000 kilometers (km) every three months, on average. Generally, whereas road inventory data are typically collected in a one-off exercise and are verified or updated when changes are made to the road (every five years or so), pavement condition data are collected at different frequencies, depending on the road class, with main roads and major highways monitored at frequent intervals (every one to two years) and minor roads assessed less frequently (at two- to five-year intervals) (Bennett et al. 2007).

The overall length of the classified road network in the extended East Africa region (map 5.1), including highways and primary, secondary, tertiary, and local roads, is estimated to be more than 1 million km according to spatial data available from the Global Roads Inventory Project for 2018. The total length of railway lines exceeds 22,000 route-km according to the World Bank's World Development Indicators, including more than 3,000 route-km each in countries such as the Democratic Republic of Congo, Mozambique, and Sudan. In addition, thousands of bridges across the region need to be inspected regularly—if one were to follow the guidelines of the US Federal Highway Administration, every two years (Gillins, Parrish, and Gillins 2016).

The demand for topographic surveys and inspections of oil and gas facilities and associated infrastructure will increase significantly in the years to come. Several large oil and gas producers are operating in African countries, and the sector continues to grow. In the power transmission sector, the total length of

MAP 5.1

Classified road network in East Africa

Source: Data from World Bank Data Catalog; Global Roads Inventory Project.
Note: Map includes highways and primary, secondary, tertiary, and local roads.

existing and planned electricity transmission lines in the countries shown in map 5.2 is more than 80,000 km, ranging from about 300 km in Djibouti to more than 12,200 km in Mozambique, according to analysis of spatial data available from the World Bank and the Africa Electricity Grids Explorer.[1] In Southern Africa, South Africa's public electric utility, Eskom, manages an impressive amount of transmission lines (more than 35,200 km route length) and towers (85,000) that require regular inspection.[2]

Even the clean energy sector has significant infrastructure inspection needs. As of the end of 2017, nearly 345,000 wind turbines were operational in the world, with more than a million turbine blades subject to wear and requiring

MAP 5.2
Electricity transmission lines in East Africa (existing and planned)

Source: Data from World Bank Data Catalog; Africa Electricity Grids Explorer.

regular inspections and care throughout their lifespans. Navigant Research predicts the global market for wind-turbine inspection by drone to grow to $6 billion by 2024 (Froese 2018). In Sub-Saharan Africa, South Africa is ramping up its wind power production capacity given its outstanding conditions for generating wind energy—more than 80 percent of South Africa's land mass has the wind conditions to produce high load factors (greater than 30 percent). As of 2019, it had 22 fully operational wind energy production facilities with 910 wind turbines in place, and another 11 facilities were under construction (SAWEA 2019). The largest wind farm in Kenya, the Lake Turkana Wind Power Project, covers an area of 160 square kilometers (km^2) and comprises 365 wind turbines (Muchira 2018). The country plans for wind power to make up 9 percent of its total energy capacity by 2030 (Kathambi Kianji 2012). With financing from the International Finance Corporation, Tanzania was set to launch its first wind farm project in 2019 in the Makambako District in the country's southwest; it is expected to have total installed capacity of 300 megawatts and about 100 wind turbines (Kemboi 2019).

PRESENT COSTS AND MODALITIES

Road pavement inspections can be time-consuming and subjective depending on the methods used. In road pavement inspection, three broad methods are used to identify distress: manual, semi-autonomous, and autonomous. Manual distress inspections are typically performed on site by workers aided by mechanical distress measurement devices (a hand odometer, a straight edge, and a ruler to measure distress length, depth, and area), requiring between 20 and 40 minutes for a crew comprising two workers for a single spot (Shaghlil and Khalafallah 2018). In many countries, evaluations of transportation infrastructure characteristics, such as road roughness, texture, mechanical and structural properties, and surface distress (from potholing and cracking to surface deformations), are still implemented primarily using manual assessments—visual or assisted by specialized equipment. Visual inspections typically suffer from human error and subjectivity (Kovacevic et al. 2016).

Establishing the financial and time costs of road network condition assessments is challenging because they tend to vary by the level of information, the level of the road network, and the technology used. Assessment of a tertiary road network could be as simple as an engineer driving along once or twice a year and making a visual inspection. Usually, a road agency or authority has a department that collects and collates the data and enters them into an asset management system to run the necessary reports for annual or multiyear planning. The time required for visual inspections tends to average about a day for every 25–30 km in East Africa, according to Workman (2018). The cost components of a visual inspection include staff wages, travel and per diem costs, vehicle and fuel (plus driver), and any equipment used, such as a GPS or a roughness measurement device. In East Africa, road agencies predominantly use visual inspection and field visits to assess the road network. In the Democratic Republic of Congo, for example, the road agency sometimes makes use of and extrapolates information received directly from road users because of budgetary and other constraints. The road agency still does force account works,[3] with road brigade teams scattered across the country that are in charge of road inspections, subject to budget availability. Data on the exact costs of inspections for the Democratic Republic of Congo in particular were not available for the current study. The assessment takes about one month, including a field visit and desk review. In the United States, pavement inspection of the road network using visual methods is estimated to cost about $17.3 per km (Congress 2018).

Available low-cost automatic data collection systems for road inspection can have a high cost-to-performance ratio, and accuracy appears to drop with road condition. Semi-autonomous inspection is composed of an automated system that collects distress data that are later classified by an expert. Existing fully automated systems for distress detection use methods such as obtaining data from specially equipped trucks traveling at freeway speeds, with resulting accuracy of as much as 95 percent; however, these systems still require a human driver. A driver is also needed for automatic data collection through low-cost smartphone applications that measure the International Roughness Index; moreover, as found in recent trials in East Africa (Kenya, Uganda, Zambia), the accuracy of results from these types of applications, such as RoadLab, tend to decline significantly on fair and poor condition unpaved roads (Workman 2018). Still, in a comprehensive guide on the topic, Bennett et al. (2007) find that, although many countries continue to use manual systems to collect data on road conditions, low-cost automated technologies can have a lower cost-to-performance ratio than manual technologies, given that ensuring quality with manual techniques is very difficult: investment of as little as $10,000 can provide road agencies with objectively quantified data.

Although remote sensing systems such as light detection and ranging LiDAR are accurate, they are expensive and difficult to integrate into routine asset management programs. For pavement evaluation, the terrestrial laser scanning and pavement profile scanner approaches using LiDAR techniques can also be used; these are vehicle-mounted systems that rely on laser scanners to construct high-resolution, continuous transverse pavement profiles. However, even though LiDAR scanners have been shown to successfully detect damage, such as distortions, rutting, shoving, and potholes, at highway speeds of up to 100 kilometers per hour (km/h), and can record measurements with very high accuracy of up to 0.30 millimeter, their measurement range limits their utility to roads up to 4 meters wide, which may be insufficient in large cities or on intercity highways (Petkova 2016). This method is also associated with high equipment and maintenance costs—and the need for trained technicians—which may make its implementation cost prohibitive and difficult to integrate into road management programs (Schnebele et al. 2015).

Finally, although aerial imagery acquired by manned aircraft and satellites can provide a valuable data source for performing pavement inspections, the limited maneuverability of airplane and satellite platforms to acquire the image data can present significant challenges to widespread usage of such platforms, particularly in urban settings (Zhang 2009). However, as concluded by the Africa Community Access Partnership project that explored different innovative solutions in the future of road management in Africa, specifically in the context of low-volume rural roads, very high resolution satellite imagery, purchased at £10.50 per square meter (approximately $13.70),[4] provides a reasonable level of accuracy in road condition assessment (between 80 percent and 87 percent for the three-level condition assessment used in Ghana, and between 63 percent and 69 percent for the five-level condition assessment used in Kenya, Uganda, and Zambia). Moreover, the assessment can be performed more quickly, at a rate of about 50–80 km per day, or even faster with increased practice, and with as much as seven times less manpower than traditional means. Satellite imagery–based assessments also permit areas to be covered that may be inaccessible because of conflict or other reasons. Workman (2018) finds that this approach, although more expensive than traditional assessments (albeit only slightly in some cases), could be cost-effective if significant discounts or relaxations on the minimum requirements for procurement of imagery could be negotiated. The approach can be particularly cost-effective in countries such as Uganda where road density, and, thus, density of roads per km^2 of imagery, is high.

Various free, open-source, and crowd-sourced tools and data, such as OpenStreetMap, Mapillary, Google Earth Engine, and others, are becoming available that can complement traditional infrastructure inventory and inspection methods. Specialized private companies have started to offer road geometry assessment and verification services using OpenStreetMap data, cross-verifying this data using satellite imagery. Some machine learning–based assessments also leverage so-called big data (for example, large volumes of data received from road user call detail records) to infer traffic flows along the network and, thus, possible issues with road conditions.

Routine inspections of bridges are regularly scheduled and consist of enough observations and measurements to determine the structural and functional condition of the bridge. In comparison, scheduled in-depth inspections or unscheduled close-up inspections of bridges are used to assess structural damage resulting from external causes and aim to detect any deficiencies not readily visible in routine inspections. In East Africa, as in most developing regions, visual inspections are still the main, if not the only, method for collecting in-service bridge condition data, with inspector safety and inspection accessibility being major concerns (Bennett et al. 2007). Inspection of critical structural components and hot spots that are hard to reach is performed mostly by special trained staff such as industrial climbers and using large under-bridge inspection units and elevating platforms. These specialized structures and the use of specially trained machine operators are associated with high logistics and personnel costs (Hallermann and Morgenthal 2013). A single under-bridge inspection vehicle required for traditional inspections costs between $500,000 and $1 million, and operational costs range from $2,000 to $3,500 per day, in addition to costs associated with lane or shoulder closures to accommodate the equipment of between $500 and $2,500 per day (UAV Expert News 2019).

Railway inspection and monitoring are considered a crucial aspect of the system and are typically performed by human inspectors, although in recent years several prototypes of vision-based inspection systems have been proposed; most have various vision sensors mounted on locomotives or wagons (Banić et al. 2019). In areas with steep slopes, visual inspection of the terrain surrounding the railway lines, as in geodetic data collection with classical methods, can result in incomplete and insufficiently detailed data, thus posing a risk to the railroad (USDOT 2018).

Most energy companies use geographic information systems to record the location of their assets (for example, power poles), from which power line information can be inferred; however, in general, the accuracy of such information is suitable only as a general guide (Li et al. 2008). Power transmission and distribution companies must also conduct periodic inspections of power line infrastructure to ensure reliable electric power distribution; the conventional methods of deploying ground staff or low-flying helicopters to complete the inspections tend to be associated with significant costs and pose health and safety concerns (Terra Drone 2019a). For example, the operating cost associated with helicopter monitoring may exceed $1,000 per hour and may expose the inspectors and the public to danger, given that the inspection process often requires low-altitude flying (SEECO 2015). For renewable energy generation infrastructure, such as wind turbines, the commonly used approaches to inspection are through telescopic lenses, or by lift or climbing (including during maintenance and repair).

INTEGRATING DRONES INTO EXISTING INFRASTRUCTURE INSPECTION SYSTEMS

How drones can add value

Industries using unmanned aircraft systems (UASs) for inspection and monitoring were projected to grow at a compound annual growth rate of more than 36 percent between 2015 and 2020, and to reach a global market size of $1.23 billion in 2020 (USDOT 2018). Drones and the data they provide are a game changer over the entire life cycle of a transport infrastructure investment, from preconstruction and construction to the operational phase; globally, the estimated total addressable value of drone-powered solutions in transport infrastructure maintenance alone is as high as $4 billion (PwC 2017). The recent advances in unmanned aerial vehicle (UAV) technologies could be key to scaling up civil infrastructure inspection with lower labor needs and costs; UAVs associated with sensing capability and computer vision can also be an innovative approach to large-scale infrastructure monitoring (Kim, Sim, and Cho 2015). In the roads sector, drones equipped with infrared sensors can be used to conduct thermal imaging; thermal maps can be used to detect structural damage to roads (Soesilo et al. 2016). Drones may hold especially large potential for the assessment of paved roads, for which satellite imagery has been found to be less useful. Also some road agencies in East Africa, such as in Uganda, are now procuring their own UAVs (Workman 2018).

However, when assessing whether and where UAVs can add value to the road inspection process, as with any road management system, the commitment is not for a one-time needs survey; rather, implementation of a road management process is a commitment to a permanent change in the way roads are managed, in which data are not only collected but can also be regularly updated (Bennett et al. 2007).

UAVs can also be used for a range of tasks in infrastructure construction management, such as to monitor progress on construction sites at various time intervals, even daily if necessary (Lin, Han, and Golparvar-Fard 2015). In a project in Haiti, for example, construction of the 10.8-km long Plaisance–Camp Coq highway was monitored using UAVs. With only four flights of 25 minutes each required to cover the construction area, the entire mission was executed over five days. Opportunities for remote monitoring of infrastructure projects become yet more important in the context of conditions such as those created by the 2020 global health pandemic.

The two primary UAV types used in the railways sector are rotary-wing UAVs, which share many characteristics with manned helicopters, including vertical takeoff and landing capability, and fixed-wing UAVs, which travel at higher speeds and over longer distances and have minimal maintenance and repair requirements. The typical application of UAVs consists of using one or several fixed-wing aircraft equipped with payloads of cameras, LiDAR and so on to fly at low altitude over a long-range section of railway (50–100 km) and at regular intervals such as once a week (Bertrand et al. 2017). The European Drones Outlook Study (SESAR 2016) predicts that railway inspection will be increasingly carried out with long-range surveying (primarily drones flying beyond visual line of sight), with the ability to monitor 200,000 km of railway track on a bimonthly basis.

Finally, UAVs, in particular, multicopters that can hover in place and are capable of vertical takeoff and landing, have been demonstrated to be useful in the inspection of bridges, enabling remote production of close-up, high-resolution still and video imagery from multiple viewing angles, with results comparable to what can be observed visually by an inspector at arm's length. Because UAVs enable bridge inspectors to remotely view the critical elements of the bridge without having to mobilize bucket trucks or close lanes of traffic, the potential safety gains and cost reductions are significant. Moreover, the comparatively low cost of operating UAVs allows frequent flights to be performed in the same area, enabling time series–based monitoring of structural health (Gillins, Parrish, and Gillins 2016).

Spatial information captured from optical remote sensors onboard UAVs has great potential for the automatic surveillance of electrical power infrastructure; overhead power line inspection in remote and rural areas is an ideal application for UAVs because of low population density and widespread power distribution (Li et al. 2008). Moreover, new UAV- and artificial intelligence–based solutions have recently been developed for the maintenance of power transmission and distribution equipment whereby the acquired data are automatically processed and analyzed by algorithms that are trained to detect crossovers at the bottom of transmission lines, buildings, and construction machinery. The estimated error (identified anomaly) detection system is accurate up to 92.5 percent (Terra Drone 2019a). For wind turbines, many of which have dimensions of 100 meters or more, an autonomous or remotely controlled UAV could be able to approach the inspection target closely with high accuracy compared with telescopic photography (Stokkeland, Klausen, and Johansen 2015).

Despite the potential, drones are still a relatively novel technology in the infrastructure inspection industry. In Sub-Saharan Africa, UAVs have very seldom been applied in this use case. One example is the Z-Roads project in Zanzibar, where, through a collaboration between the UK Department for International Development and consultants based at the University of Nottingham, drone imagery–based surveying and state-of-the-art machine learning were used to assess the condition of low-volume roads (N/LAB and Digital Economy Consultants 2018). The survey covered all roads under the responsibility of the Department of Roads in Zanzibar, approximately 700 km of roadway.

Terra Drone Corporation, one of the world's largest industrial drone service companies, provides industrial drone technologies empowered with LiDAR and photogrammetric surveying methods for the construction, electricity, energy, and oil and gas sectors. In 2019, the company established a permanent presence in Angola after receiving multiple contracts from major oil and gas companies in West Africa. As of late 2019, the company had inspected more than 90,000 km of power lines throughout the world using beyond visual line of sight technologies (Terra Drone 2019b). In East Africa, the use of drones for inspecting energy infrastructure—transmission lines, distribution centers, and towers—is starting to be scaled up in Rwanda, with local drone operators such as Charis providing the service on a contract basis to the national Ministry of Infrastructure. As reported by Charis, the initiative has moved from a pilot into a full-time mode, with the demand for transmission line inspection spanning the entire country and that for distribution station inspection focused on certain localities.

Finally, drones have also been used to survey port facilities and monitor ongoing construction on the African continent, such as in Casablanca, Morocco (African Union and NEPAD 2018). Regular flights using multirotor drones flying at low altitudes are organized over the port to ensure contractors meet their deadlines. High-definition images that are captured by airborne sensors are used by port officials for monitoring and eventually seeking further information about or launching investigations into the progress of ongoing construction.

In sum, the main applications of UAVs in transport infrastructure inspection and management include the following:

- Road identification and inventory (presence and location of road segments, connectivity between road segments, road geometry)
- Topographic surveys during road infrastructure construction and works
- Regular surveys and monitoring to assess condition of road, bridge, and railway infrastructure

In the energy and extractives sector, the most common applications of UAVs include the following:

- Power line detection (combining UAV imagery with algorithms for automatic power line extraction)
- Power line, pipeline, and wind turbine inspections
- Power infrastructure damage assessment (such as following storms)
- Mine planning, and 2D and 3D surveying of existing mines; quick volume calculations
- Offshore oil and gas rig inspection
- Solar photovoltaic panel and concentrated solar power plant inspection

Potential time and cost savings and quality improvements

Capital investment in highway and road infrastructure maintenance is critical to economic growth and prosperity; high-quality inspection and timely maintenance of such assets are crucial preventive measures against their failure or subsequent expensive repair that may be costly. Drone technology enables quality expectations to be met while often reducing the associated cost and time; for example, infrastructure construction sites, such as those for roads and highways, can be surveyed up to 20 times faster by drones than by ground-based land surveying teams (PwC 2017). Numerous studies have explored, and largely confirmed, the quality advantages of using drones in road condition assessments, as in Zhang (2006) and Zhang and Elaksher (2012). Zhang and Elaksher (2012) used UAVs specifically to measure distress of unpaved road surfaces. The comparison of the derived 3D information with the on-site manual measurement of road distress suggests that the drone-based assessments are accurate up to 0.5 cm.

Drones have been used in bridge inspection in the United States for several years, and have proven effective in improving the quality of inspections, reducing costs, and enhancing safety. The Minnesota Department of Transportation, for example, found that drone-assisted bridge inspections saved at least 40 percent of the cost of their traditional counterparts, mainly from savings on traffic control and equipment, and that the inspection could be completed in five days compared with eight for traditional methods. A cost-benefit analysis for the Oregon Department of Transportation showed a cost-benefit ratio of 9 for conducting bridge inspections using UAVs, with average savings of $10,000 per bridge inspection (Gillins et al. 2018). As of 2018, 35 state departments of transportation were using drones (Slowey 2019).

In a US-based study, Brooks (2016) estimates that the cost of pothole detection using UASs amounts to $0.80 per km, which is a fraction of the cost associated with ground-based collection methods (up to $174 per km). Surveys conducted by Congress (2018) suggest that the cost for bridge and pavement inspection services offered by US UAV companies that charge daily rates range from $2,000 to $5,000 per day, in addition to a per diem charge of $150–$250 if the work period exceeds eight hours. Such companies tend to be capable of complex operations, such as under-bridge inspections. Companies that use hourly rates charge from $150 to $350 per hour. In comparison, the rates for visual observers (US specialists) generally range from $85 to $300 per hour, in addition to further personnel costs associated with traffic control enforcement, which can be avoided if using UAVs. In fact, the most important overall cost savings of UAV-assisted inspections compared with visual inspections may be due to lessened traffic congestion: road closures and time during which traffic is impaired can be limited, which is particularly important for high-traffic bridges (Dorafshan and Maguire 2018). UAV equipment costs using the Leica P20 scanner are only a fraction of terrestrial laser scanning bridge inspection methods, less than 3 percent ($2,500 vs. $103,000), and the on-site survey time is about two-thirds less (1 hour vs. 3 hours) (Chen et al. 2019). Compared with manned aircraft–based data collection, the use of fixed-wing drones such as the eBee can help reduce costs by 2 to 10 times.

Research by Shaghlil and Khalafallah (2018) demonstrates that road pavement distress could be satisfactorily detected using a DJI Mavic Pro Quadcopter, which costs about $1,100, equipped with a regular smartphone (at additional cost) to capture images and video, flying at a height between 5 and 25 meters. Similarly, Gillins, Parrish, and Gillins (2016) show that imagery and high-definition video footage that can be used by bridge inspectors to identify potential defects that need attention could be captured using low-cost, camera-equipped multicopter drones such as the DJI Phantom 3 Pro multicopter, which retails for about $1,000 when bundled with a 4K video camera. Despite these promising findings, the price of a UAV for bridge inspection varies significantly, depending on the purpose of the inspection, quality and quantity of the integrated sensors, and computing capabilities (Dorafshan and Maguire 2018). For example, integrating thermal cameras with existing visual sensors can increase the price of the UAS up to three times (to more than $5,500 for a drone such as the DJI Phantom 4 Pro); the size and the price of the UAV increase yet more dramatically (to $30,000 and more) if autonomous 3D model reconstruction is a requirement of the inspection. Recent research on the use of UAVs in rural road inspection in East African countries also indicates that the manpower-associated costs of UAVs will be significantly increased if restrictions are in place on the times and situations in which drones can be flown, such as not over the traffic or not during certain times of the day (see Workman 2018).

Comparative cost evaluations conducted by Workman (2018) also provide insights into price parity of drone-based rural roads inspections in individual East African countries, given the costs associated with traditional assessments and satellite assessments, which are the two main alternatives. As shown in figure 5.1, the cost per km of road inspected is highest in Zambia, where traditional assessment and satellite imagery–based assessment both cost about $39–$40 per km. Parity is by far the lowest in Uganda, where, partly because of the country's very high road density, traditional assessment and satellite imagery–based inspection cost $12 per km and $13.3 per km, respectively.

FIGURE 5.1

Cost of rural road inspection using traditional assessment methods and satellite image–based assessment

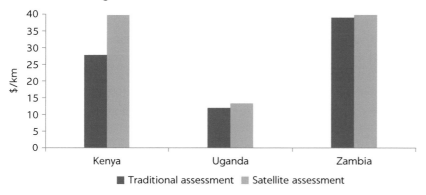

Source: Based on data from Workman (2018).

It should be noted, however, that the satellite imagery–based cost excludes the cost of training facilities.

UAV technology is promising with regard to the time required for infrastructure inspection; any cost savings from using UAVs for road network and bridge inspection may derive not so much from lower hourly rates but from the aggregate costs, which are a function of the time required for each inspection. Civilian eBee drones, for example, reach a cruise speed of 50 km/h, with wind resistance of up to 46 km/h, and thus have a speed advantage over other methods. Drones can perform assessments of rail and road networks four to five times faster than terrestrial survey methods. For the Z-Roads project in Zanzibar, the survey of the 700-km-long road network was completed in 12 days by a project team that included the Department of Roads and the private companies N/LAB and Spatial Info. However, as with costs, there are caveats: if automated or semi-automated tools are not available to the specific transport agency, drone image postprocessing can require additional time to perform as well as appropriately skilled manpower.

Yang et al. (2015) find that it is feasible to undertake a survey of the structure of an entire bridge (240 meters long) within 42 minutes using a UAS, including setup time. For a large-scale bridge (2,400 meters), analysis by Wells and Lovelace (2017) shows that UAS inspection was 37 percent faster and 66 percent cheaper than traditional inspection; however, details on the comparative inspection performance were not reported. In under-bridge inspections, the largest cost component associated with visual inspections, slightly more than 50 percent of the total cost, is typically the deployment of under-bridge inspection trucks, followed by the cost of inspectors and the cost of traffic control. With UAV inspections, no under-bridge inspection trucks are necessary, and the other two cost components are significantly reduced (Chan et al. 2015).

In railway line inspection, including regular safety checks for wear and tear, vegetation management, rock falls, security, and post-weather-event evaluations, drones can significantly reduce the amount of time required. Substantial cost savings associated with using drones compared with traditional inspection methods are mainly due to elimination of the need for traffic shutdowns during inspections and lower labor costs. Because railway embankments are linear structures that can be dozens and even hundreds of kilometers long, visual inspection by personnel on the ground may cause very important information to be missed, especially if the track is located on high or steep embankments. UAVs can enable higher-quality visual assessments to be made, although certain physical parameters of the topography, such as subsoil composition, may not be detected (Kovacevic et al. 2016). Using drones for computer vision–based monitoring of railway infrastructure has been demonstrated to be a cost-effective, robust, and reliable visual inspection method (Banić et al. 2019).

A major advantage of using drones in the oil and gas industry is that unmanned technologies eliminate the need for any scaffolding and rope access or even costly shutdowns of processing plants (Terra Drone 2019b). The global oil and gas industry, which manages 10 million km of pipelines worldwide, spends $37 billion a year monitoring above- and below-ground pipes. It is estimated that drones can reduce traditional surveillance costs by nearly 90 percent, rapidly survey hundreds of km of pipeline, and produce aerial imagery that pinpoints landscape changes that indicate pipeline leaks (Wall 2019).

Drones are also used in many aspects of utility operations, including power line inspections and repair, with significant cost savings, given that they can fly relatively close to the power line, providing a cheap and flexible way to gather spatial data from the power line corridor (Li et al. 2008). An early evaluation of UAVs for power line inspection indicated that this approach could be faster than foot patrol and would yield the same or even better accuracy than costly helicopter inspection (Jones 1996). For example, during restoration work in Puerto Rico following Hurricane Maria, a team used drones to pull 72,000 feet of rope for conductor wire, requiring just eight weeks to complete work that would have taken six to eight months using traditional methods (UAV Expert News 2019).

Some of the drone companies operating in the power line and tower inspection space in Africa are still in search of a sustainable business model, noting that funding themselves has been one of the major challenges while they engage with and convince utilities to become their clients and regularly use their services. One locally operating drone company reports that an issue in South Africa has been the South African Civil Aviation Authority's application process for a commercial operating license, which is expensive and lengthy and has caused many companies to run out of money before completing it.[5] However, the economic and practical benefits of using drone technology in power transmission and distribution infrastructure inspection have been demonstrated in Rwanda, for example, where drones are starting to replace manual inspections. According to the interviewed local drone operators, drone-based inspections have reduced the time requirements significantly: a team is able to inspect as many as 30 towers per day, compared with two or three towers per day using manual inspection. Moreover, with drones, energy infrastructure in parts of the country that would have been completely inaccessible to manual inspectors can now be inspected.

Drone inspections of turbine towers and blades can save a wind farm owner time and money and reduce safety risks compared with conventional, rope-based inspections. Some of the economic benefits of using drone solutions for visual inspection of on- and offshore wind turbines include a reduction in wind turbine downtime and the cost savings from using an entirely digital platform. UAV-based inspections enable wind project owners and operators to avoid sending human inspectors down on a rope or up in a basket on a crane to inspect blades up close—work that is even more expensive and dangerous for turbines located in remote locations far offshore (Gerdes 2019). Wind turbine inspection drone companies suggest that, compared with ground-based methods, drones can lower the inspection costs by 30–50 percent (and enable the wind farm to remain operational during inspection) while also increasing the inspection resolution.[6] Still, the costs of UAV-based inspections are not low: wind farm owners and operators globally currently spend up to $25,000 per drone to be able to safely operate the drones in high winds. Consumer-grade drones, although available at a fraction of the cost of commercial drones, also tend to lack reliable remote-control systems, camera quality for high-resolution imaging, and software platforms for data collection and analysis. Project owners, especially those owning multiple wind sites, typically prefer to use their own drones rather than turbine-blade inspection company services (Froese 2018).

In the solar energy use case, drones cut the cost of thermographic inspections of utility-scale solar farms by 30–40 percent, reducing the time required to inspect a 75-megawatt plant from about a month to about a week (Wall 2019). The cost reported by specialized drone companies for thermographic photovoltaic panel inspections is between $2,000 and $3,000 a day, covering two hectares in an hour (UAS Vision 2016). According to pilot tests, an eBee drone can inspect a site composed of 4,000 solar panels in about five minutes, whereas a human inspection would take more than eight days at a rate of one panel inspection per minute (Wade 2016).

THE HUMAN AND SOCIETAL IMPACTS

Infrastructure data collection is expensive; therefore, road and other transport agencies must only collect the data required for management purposes, but that data must be collected at a frequency and a level appropriate for the decisions they are to be used for (Bennett et al. 2007). Regular and affordable UAV-enabled infrastructure inspection and monitoring can save limited public funds; improve budgetary planning at the city, regional, and national levels; and help ensure the overall ability of the infrastructure network to efficiently serve the needs of local and regional economies.

In addition to these economic implications, individual UAV use cases in the infrastructure space can also affect human life and well-being. In infrastructure construction management, these applications range from the detection and prevention of trespassing to placement of trench protection to enhancing site safety. In one example, it was estimated that the number of life-threatening accidents on an average construction site monitored by drones has been decreased by up to 91 percent (PwC 2017). Civil infrastructure, including bridges and power plants, is exposed to loadings such as earthquakes and ocean waves that may cause structural deterioration and damage; inability to detect the damage in a timely manner, in turn, can cause catastrophic collapses associated with significant socioeconomic and human losses. UAV service and artificial intelligence inspection of power transmission lines and other infrastructure can reduce health and safety concerns by eliminating low-altitude manned helicopter missions and reducing the number of workers in the field.

Finally, compared with traditional visual infrastructure inspection methods, drone-based approaches can also potentially reduce the environmental footprint. In Tanzania, for example, an estimated 30,000 or more vehicle-kilometers would have to be driven to inspect the 110,000 km of rural roads if smaller roads are driven twice (there and back). The associated emissions of a single complete inspection would amount to about 130 tons of carbon dioxide (Workman 2018). Although the carbon footprint of UAV-based inspections is less certain and depends on the exact UAV model and its technical specifications, UAV inspections would necessarily avoid a significant portion of the fossil fuel use associated with the traditional approach.[7]

NOTES

1. "Africa Electricity Grids Explorer," http://africagrid.energydata.info/.
2. Profile at the end of financial year 2016/17, Scopito, https://scopito.com/rise-of-drone-inspections-in-africa/.
3. Force account work, also known as work-by-force account, or time and materials work, is a payment method for construction work in which there is no existing agreement on cost.
4. A total of 1,829 km^2 of imagery was acquired for the project to assess segments of rural roads in Ghana, Kenya, Uganda, and Zambia. Other costs associated with satellite imagery–based assessments include the cost of the calibration, based on traditional surveys, and staff wages to assess the imagery and produce results.
5. Scopito, https://scopito.com/rise-of-drone-inspections-in-africa/.
6. "Drone Wind Turbine and Blade Inspection for Offshore and Onshore Wind Farms," https://abjdrones.com/drone-wind-turbine-inspection/.
7. Drone batteries will be charged by the electric grids where they are operating, which differ by region and over time in their fuel mixes and corresponding emissions (Stolaroff et al. 2018).

REFERENCES

African Union and NEPAD (New Partnership for Africa's Development). 2018. "Drones on the Horizon: Transforming Africa's Agriculture." NEPAD, Midrand, South Africa.

Banić, M., A. Miltenović, M. Pavlović, and I. Ćirić. 2019. "Intelligent Machine Vision Based Railway Infrastructure Inspection and Monitoring Using UAV." *Mechanical Engineering* 17 (3): 357–64.

Bennett, C. R., A. Chamorro, C. Chen, H. de Solminihac, and G. W. Flintsch. 2007. "Data Collection Technologies for Road Management." Version 2.0. World Bank, Washington, DC.

Bertrand, S., N. Raballand, F. Viguier, and F. Muller. 2017. "Ground Risk Assessment for Long-Range Inspection Missions of Railways by UAVs." Paper prepared for International Conference on Unmanned Aircraft Systems (ICUAS), Miami, FL, June 13–16. doi:10.1109/ICUAS.2017.7991331.

Brooks, C. N. 2016. "Applications of UAVs for Transportation Infrastructure Assessment." Michigan Tech Research Institute, Ann Arbor, MI.

Chan, B., H. Guan, J. Jo, and M. Blumenstein. 2015. "Towards UAV-Based Bridge Inspection Systems: A Review and an Application Perspective." *Structural Monitoring and Maintenance* 2 (3): 283–300. doi:10.12989/smm.2015.2.3.283.

Chen, S., D. F. Laefer, E. Mangina, I. Zolanvari, and J. Byrne. 2019. "UAV Bridge Inspection through Evaluated 3D Reconstructions." *Journal of Bridge Engineering* 24 (4).

Congress, S. S. C. 2018. "Novel Infrastructure Monitoring Using Multifaceted Unmanned Aerial Vehicle Systems–Close Range Photogrammetry (UAV-CRP) Data Analysis." Dissertation presented to the Faculty of the Graduate School of The University of Texas at Arlington in Partial Fulfillment of the Requirements for the Degree of Doctor of Philosophy.

Dorafshan, S., and M. Maguire. 2018. "Bridge Inspection: Human Performance, Unmanned Aerial Systems and Automation." *Journal of Civil Structural Health Monitoring* 8 (3): 443–76.

Froese, M. 2018. "How to Choose the Right Drone to Inspect Your Wind Turbines." Windpower Engineering & Development, August 23. https://www.windpowerengineering.com/how-to-choose-the-right-drone-to-inspect-your-wind-turbines/.

Gerdes, J. 2019. "Drones and Crawling Robots Will Soon Be Inspecting Wind Turbines." *Green Tech Media*, July 22. https://www.greentechmedia.com/articles/read/drones-and-crawling-robots-will-soon-be-inspecting-wind-turbines.

Gillins, D. T., C. Parrish, M. N. Gillins, and C. Simpson. 2018. "Eyes in the Sky: Bridge Inspections with Unmanned Aerial Vehicles." Oregon State University, Corvallis, OR, for Oregon Department of Transportation.

Gillins, M. N., C. Parrish, and D. T. Gillins. 2016. "Cost-Effective Bridge Safety Inspections Using Unmanned Aircraft Systems." Conference Paper, Geotechnical and Structural Engineering Congress, Phoenix, AZ, February 14–17. doi:10.1061/9780784479742.

Hallermann, N., and G. Morgenthal. 2013. "Unmanned Aerial Vehicles (UAV) for the Assessment of Existing Structures." Conference Paper, IABSE Symposium 2013, Kolkata. doi:10.2749/222137813808627172.

Jones, D. I. 1996. *Requirements for Aerial Inspection of Overhead Electrical Power Lines.* Chester, U.K.: EA Technology. https://books.google.com/books/about/Requirements_for_Aerial_Inspection_of_Ov.html?id=U8TdGwAACAAJ.

Kathambi Kianji, C. 2012. "Kenya's Energy Demand and the Role of Nuclear Energy in Future Energy Generation Mix." Paper presented at the Joint JAPAN-IAEA Nuclear Energy Management School, Tokaimura, Japan, June 11–29. https://web.archive.org/web/20130613001823/http://www.iaea.org/nuclearenergy/nuclearknowledge/schools/NEM-school/2012/Japan/PDFs/week2/CR6_Kenya.pdf.

Kemboi, L. 2019. "Tanzania Set to Launch Its First Wind Power Project." *Construction Review Online*, March 13. https://constructionreviewonline.com/2019/03/tanzania-set-to-launch-its-first-wind-power-project/.

Kim, H., S. H. Sim, and S. Cho. 2015. "Unmanned Aerial Vehicle (UAV)-Powered Concrete Crack Detection Based on Digital Image Processing." Paper prepared for 6th International Conference on Advances in Experimental Structural Engineering, University of Illinois, Urbana-Champaign, August 1–2.

Kovacevic, M. S., K. Gavin, I. Stipanovic Oslakovic, and M. Bacic. 2016. "A New Methodology for Assessment of Railway Infrastructure Condition." *Transportation Research Procedia* 14: 1930–39.

Li., Z., Y. Liu, R. Hayward, J. Zhang, and J. Cai. 2008. "Knowledge-Based Power Line Detection for UAV Surveillance and Inspection Systems." Paper prepared for 23rd International Conference Image and Vision Computing, New Zealand, November 26–28. doi:10.1109/IVCNZ.2008.4762118.

Lin, J. J., K. K. Han, and M. Golparvar-Fard. 2015. "A Framework for Model-Driven Acquisition and Analytics of Visual Data Using UAVs for Automated Construction Progress Monitoring." Paper prepared for 2015 International Workshop on Computing in Civil Engineering, Austin, TX, June 21–23. https://doi.org/10.1061/9780784479247.020.

Muchira, N. 2018. "Lake Turkana Wind Power $52.5m Fine Pushed to Consumers." *The East African*, October 27. https://www.theeastafrican.co.ke/business/Lake-Turkana-Wind-Power-fine-pushed-to-consumers-/2560-4825152-vff05t/index.html.

N/LAB and Digital Economy Consultants. 2018. "Z-ROADS: Drone Imagery and Applied Machine Learning for the Assessment of Low Volume Road Condition Analysis." Presented at Lake Victoria Challenge, Mwanza, Tanzania, October 29–31.

Petkova, M. 2016. "Deploying Drones for Autonomous Detection of Pavement Distress." Submitted to the Program in Media Arts and Sciences, School of Architecture and Planning, in partial fulfillment of the requirements for the degree of Master of Science at the Massachusetts Institute of Technology.

PwC (PricewaterhouseCoopers). 2017. "Clarity from above: PwC Global Report on the Commercial Applications of Drone Technology." PricewaterhouseCoopers, Warsaw.

SAWEA (South Africa Wind Energy Association). 2019. "South African Wind Farms." South Africa Wind Energy Association. https://sawea.org.za/wind-map/wind-ipp-table/.

Schnebele, E., B. F. Tanyu, G. Cervone, and N. Waters. 2015. "Review of Remote Sensing Methodologies for Pavement Management and Assessment." *European Transport Research Review* 7 (2). doi:10.1007/s12544-015-0156-6.

SEECO (Southern Electrical Equipment Company, Inc.). 2015. "Unmanned Aerial Vehicle for Monitoring Infrastructure Assets." United States Patent No. US9,162,753 B1. Southern Electrical Equipment Company, Inc., Charlotte, NC.

SESAR (Single European Sky Air Traffic Management Research). 2016. *European Drones Outlook Study: Unlocking the Value for Europe.* Single European Sky Air Traffic Management Research. Brussels: European Commission.

Shaghlil, N., and A. Khalafallah. 2018. "Automating Highway Infrastructure Maintenance Using Unmanned Aerial Vehicles." Conference Paper, Construction Research Congress, New Orleans, LA, April 2–4. doi:10.1061/9780784481295.049.

Slowey, K. 2019. "More Evidence That Drones Could and Should Play Major Role in Infrastructure Inspections." Construction Dive, February 6. https://www.constructiondive.com/news/more-evidence-that-drones-could-and-should-play-major-role-in-infrastructur/547684/.

Soesilo, D., P. Meier, A. Lessard-Fontaine, J. Du Plessis, and C. Stuhlberger. 2016. "Drones in Humanitarian Action: A Guide to the Use of Airborne Systems in Humanitarian Crises." FSD (Swiss Foundation for Mine Action), Geneva.

Stokkeland, M., K. Klausen, and T. A. Johansen. 2015. "Autonomous Visual Navigation of Unmanned Aerial Vehicle for Wind Turbine Inspection." Paper presented at 2015 International Conference on Unmanned Aircraft Systems (ICUAS), Denver, CO, June 9–12. doi:10.1109/ICUAS.2015.7152389.

Stolaroff, J., C. Samaras, E. R. O'Neill, A. Lubers, A. S. Mitchell, and D. Ceperley. 2018. "Energy Use and Lifecycle Greenhouse Gas Emissions of Drones for Commercial Package Delivery." *Nature Communications* 9: 409. doi:10.1038/s41467-017-02411-5.

Terra Drone. 2019a. "Terra Drone Launches UAV AI-Based Solution for Power Asset Inspection Developed after Inspecting over 90,000 km Power Lines by BVLOS." Terra News, October 3. https://www.terra-drone.net/global/2019/10/03/terra-drone-launches-uav-ai-based-solution-for-power-asset-inspection-developed-after-inspecting-over-90000-km-power-lines-by-bvlos/.

Terra Drone. 2019b. "Terra Drone Opens Angola Branch Due to High Demand from Oil and Gas Industry." Terra News, May 10. https://www.terra-drone.net/global/2019/05/10/terra-drone-angola-demand-oil-gas-industry-africa/.

UAS Vision. 2016. "Drones Cut Cost of Thermographic PV Panel Inspections." UAS Vision, September 15. https://www.uasvision.com/2016/09/15/drones-cut-cost-of-thermographic-pv-panel-inspections/.

UAV Expert News. 2019. "Drones Are Lowering the Cost of Infrastructure Inspection." UAV Expert News, May 9. http://www.uavexpertnews.com/2019/05/drones-are-lowering-the-cost-of-infrastructure-inspection/.

USDOT (United States Department of Transportation). 2018. "Unmanned Aircraft System Applications in International Railroads." US Department of Transportation, Federal Railroad Administration, Washington, DC.

Wade, M. 2016. "Conducting a Solar Panel Inspection with an eBee Drone." Waypoint, February 22. https://waypoint.sensefly.com/conducting-a-solar-panel-inspection-with-an-ebee-drone/.

Wall, R. 2019. "8 Ways Drones Are Lowing the Cost of Infrastructure Inspection." Power Engineering. https://www.power-eng.com/2019/05/06/8-ways-drones-are-lowering-the-cost-of-infrastructure-inspection/.

Wells, J., and B. Lovelace. 2017. "Unmanned Aircraft System Bridge Inspection Demonstration Project Phase II Final Report." No. MN/RC 2017-18. Collins Engineers for Minnesota Department of Transportation, St. Paul, MN. http://dot.state.mn.us/research/reports/2017/201718.pdf.

Workman, R. 2018. "The Use of Appropriate High-Tech Solutions for Road Network and Condition Analysis, with a Focus on Satellite Imagery." Final Report by TRL Ltd for Africa Community Access Partnership and UKaid.

Yang, C. H., M. C. Wen, Y. C. Chen, and S. C. Kang. 2015. "An Optimized Unmanned Aerial System for Bridge Inspection." *Proceedings of the International Symposium on Automation and Robotics in Construction* 32: 1–6.

Zhang, C. 2006. "A UAV-Based Photogrammetric Mapping System." *International Archives of the Photogrammetry, Remote Sensing and Spatial Information Sciences* XXXVII: 627–31.

Zhang, C. 2009. "Monitoring the Condition of Unpaved Roads with Remote Sensing and Other Technology." Final Report for US DOT DTPH56-06-BAA-0002, South Dakota State University, Brookings, SD.

Zhang, C., and A. Elaksher. 2012. "An Unmanned Aerial Vehicle-Based Imaging System for 3D Measurement of Unpaved Road Surface Distresses." *Computer-Aided Civil and Infrastructure Engineering* 27 (2): 118–29. http://dx.doi.org/10.1111/j.1467-8667.2011.00727.x.

6 Other Applications

DISEASE MONITORING AND PREVENTION

Unmanned aerial vehicles (UAVs) can be used to the benefit of public health, particularly in vector control. For example, UAVs have been used in efforts to contain the spread of malaria in Cheju, Tanzania. A nontoxic, biodegradable control agent called Aquatain AMF was sprayed on rice paddies (a breeding habitat) to kill mosquito larvae. As a result of the pilot project, the mosquitoes that carry malaria were almost entirely eliminated, reducing the prevalence of malaria, which in Tanzania infects more than 10 million people every year and where eradication efforts until now have largely focused on costly and time-consuming manual spraying (Wight 2019). Data to identify mosquito breeding sites, so that the larvae can be controlled, were collected in the Kasungu District of Malawi in 2018 by a team of scientists from the Liverpool School of Tropical Medicine and Lancaster University.[1] Drones have also been used in the region in combination with artificial intelligence to identify risk zones for cholera, for example, as part of a UNICEF project in Malawi. The availability of the UAV-generated maps also led to faster responses during cholera outbreaks in Tanzania (Soesilo et al. 2016). Projects in Ethiopia to combat the tsetse fly are using the greater maneuverability of drones to improve the targeted use of sterile vectors to stem the spread of disease (USAID 2017).

PANDEMIC RESPONSE

New drone applications in the public health field have also emerged in the context of the global COVID-19 (coronavirus) pandemic. Health supply chains collapsed in countries around the globe, with demand for medical products surging amid reduced production and an inability to import such commodities. Efficient management of the reduced inventory thus has become critical. Having a drone distribution network in place—or even a drone corridor, as in Ghana and Rwanda—before the pandemic enables countries to quickly scale up a response. The low-hanging fruit, as demonstrated by operations in Ghana, would be the creation of more centralized management of critical medical products (diagnostic test kits, medicine, and personal protective equipment) and the ability to effectively dispatch these products on demand without compromising the safety of, or potentially contaminating, personnel. In China, for instance, during the peak of the pandemic, drones were used for aerial spraying and disinfection, transport of samples, e-commerce delivery, public announcements, and crowd monitoring. In the spring of 2020 in Sierra Leone, the national government used UAV technology, along with other sources of data (call detail records and the like), to monitor compliance with lockdown policies in the western part of the country. In this context in particular, the need for developing protocols and regulations to ensure that drone-based surveillance does not conflict with personal privacy and human rights principles has been increasingly acknowledged.

CONSERVATION

UAVs hold large potential to monitor national parks for illegal poaching and deforestation. They are already used in Kruger National Park in South Africa and in Namibia to fight rhino and elephant poaching. As of mid-2020, the Kenyan Park Service had plans to introduce the technology, flying thermo-drones at night, which is safer than sending rangers.

PEACE AND SECURITY

Some human rights organizations have proposed using UAVs to monitor human rights and avert atrocities, including monitoring the movements of armed groups and functioning as an early warning system or documenting evidence of war crimes. These organizations include the Sentinel Project and the Genocide Intervention Network. However, such monitoring may cause considerable anxiety among the population in contexts in which drones are also used to conduct attacks and where the population may not be able to distinguish between armed and unarmed drones (OCHA 2014). The UN Security Council approved the use of unmanned drones by the UN Organization Stabilization Mission in the Democratic Republic of the Congo (MONUSCO) in January 2013, and drones were officially launched in December of that year. MONUSCO is using UAVs to promote peace and security, including by assessing population movements, assessing environmental challenges, and conducting needs evaluations. The drones, each of which contains a camera with infrared and synthetic-aperture radar[2] capabilities, are based in Goma and are deployed across North Kivu (USAID 2017). There have also been discussions about adding UAV capacity to other UN missions, including those in Côte d'Ivoire, South Sudan, and Sudan (OCHA 2014).

MINING AND INDUSTRY

Unmanned aircraft systems (UASs) have become an important tool in the field of geoscience and can be particularly useful for extracting geological information from inaccessible areas. In the mining sector, in a single automated flight a drone can collect timely, georeferenced imagery that is quickly transformed into a precise 3D copy of the site. The digital version could be used to calculate volumes, perform site surveys, optimize traffic management, design road layouts, and so on.[3]

Aerial surveying and mapping also offer cost and other advantages in the construction and industrial sector. For example, LafargeHolcim, a leading building materials and solutions company, has a head office and fully integrated plant in the Mbeya Region in southwest Tanzania. To deal with issues managing inventory, the company has been using drones to undertake inventory surveys. According to Hossam Elzohery, plant manager, the drone surveys proved to be more accurate than normal stockpile surveys (with 250–300 pictures per 25–30-minute flight). Substantial sums were previously spent on manual surveying and bringing in outside experts. The traditional surveys took two days and the people doing the surveys were exposed to potential hazards. The drone surveys take one day and aids in the planning of targeted blasts and mine rehabilitation. It took four months for the company to break even on two eBee mining drones (SenseFly 2017).

In South Africa, UAS technology was used in opencast highwall mapping in Mpumalanga Province. The UAV flyover generated data within minutes (35 minutes over an area 500 meters × 240 meters), and the 3D model generated from the raw data showed a good correlation with the resource model (Katuruza and Birch 2019).

NOTES

1. "Drones vs Mosquitoes: Fighting Malaria in Malawi," https://www.unicef.org/stories/drones-vs-mosquitoes-fighting-malaria-malawi.
2. Synthetic-aperture radar is a type of active data collection in which a sensor produces its own energy and then records the amount of that energy reflected back after interacting with the Earth.
3. "The Professional's Mapping Drone of Choice," SenseFly, https://www.sensefly.com/.

REFERENCES

Katuruza, M., and C. Birch. 2019. "The Use of Unmanned Aircraft System Technology for Highwall Mapping at Isibonelo Colliery, South Africa." *Journal of the Southern African Institute of Mining and Metallurgy* 119: 291–95.

OCHA (United Nations Office for the Coordination of Humanitarian Affairs). 2014. "Unmanned Aerial Vehicles in Humanitarian Response." Occasional Policy Paper, United Nations Office for the Coordination of Humanitarian Affairs, New York and Istanbul.

SenseFly. 2017. "Using Drones to Optimise Cement Plant Efficiency." YouTube. September 11. https://www.youtube.com/watch?v=neQ2s87wYKc.

Soesilo, D., P. Meier, A. Lessard-Fontaine, J. Du Plessis, and C. Stuhlberger. 2016. "Drones in Humanitarian Action: A Guide to the Use of Airborne Systems in Humanitarian Crises." FSD (Swiss Foundation for Mine Action), Geneva.

USAID (United States Agency for International Development). 2017. *Unmanned Aerial Vehicles Landscape Analysis: Applications in the Development Context.* Global Health Supply Chain Program-Procurement and Supply Management. Washington, DC: Chemonics International Inc. for United States Agency for International Development.

Wight, A. 2019. ""How Do You Fight Malaria in Tanzania? With Drones!" Forbes, November 10. https://www.forbes.com/sites/andrewwight/2019/11/10/how-do-you-fight-malaria-in-tanzania-with-drones/#149262a5bab0.

7 Other Considerations in Drone Economics

OPPORTUNITIES FOR COST REDUCTION BY COMBINING USE CASES

As shown by the field research conducted for the study in Ukerewe District, Tanzania (see chapter 1), drone costs per kilometer can be reduced significantly by improving the time utilization of each unmanned aerial vehicle (UAV). However, layering use cases involves collaboration across multiple health programs, each of which may have its own stakeholders, managers, and priorities. But layering of use cases need not be limited to health. Unmanned aircraft systems (UASs) can operate as logistics providers serving a variety of sectors. One sector that is potentially promising for UAS delivery is agriculture, particularly livestock. Vaccines to protect against livestock diseases are small volume, high-value items that are difficult to store and transport because they require temperature control. Demand for them also occurs in the same rural regions where many health facilities may be located. Thus, there could be an opportunity to layer commercial use cases from *outside* the health sector for delivery by the same UAS.

Greater clarity around required versus nice-to-have UAV characteristics will be critical to enable pragmatic trade-offs between optimal UAVs and simpler versions or minimum viable products. More transparency is also needed around both manufacturing costs and willingness to pay in global health to avoid a disconnect between UAV innovators and global health customers (USAID 2018).

THE ROLE OF THE REGULATORY ENABLING ENVIRONMENT

For cost-efficient delivery, a drone should be allowed to fly in an automated manner from one location to another that can be many kilometers away, provided all safety requirements are in place. For example, in the railway infrastructure use case, the high speed and distance capabilities of fixed-wing UAVs can be best utilized when the UAVs can fly beyond visual line of sight. However, regulations in many countries currently do not permit this type of operation (Drones for Development 2016; USDOT 2018). As of 2018, UAV regulations were in place in only about a quarter of all African countries (African Union and

NEPAD 2018): South Africa was the first in the region to implement and enforce a comprehensive set of legally binding rules governing UAVs in July 2015, and 14 other countries published dedicated UAV regulations in July 2017, including Madagascar, Rwanda, Tanzania, Zambia, and Zimbabwe. Other countries had made minor changes to existing regulations, and the rest were either in the process of developing regulations for UAVs or had taken no action. Although this lack of regulation has created an opportunity for experimentation, it is clearly not a sustainable path for UAV use at scale (USAID 2018). The pace of growth of UAS use is outpacing the ability of national regulatory authorities to develop the rules and the systems needed to govern their use (UNICEF Supply Division 2019). Civil aviation authorities are traditionally risk averse and slow to evolve, and many pilots have been conducted as exemptions in the absence of regulations or policies specifically governing the use of UASs to transport public health payloads (VillageReach 2019).

The lack of regulatory clarity is seen as the largest barrier to wider adoption of UAVs in the agricultural industry as nations grapple with the problem of keeping UAVs legal while securing air safety and privacy rights. The lack of regulatory guidance discourages investment, as shown by a case in which a Rwandan agricultural drone company hesitated to expand into neighboring Uganda because of Uganda's lack of clear drone policies (CTA 2019). The drone operators in Tanzania interviewed for the current study likewise opined that the drone industry in the country is still in an early stage and therefore vulnerable to even minor regulatory changes. Although the government sees value in drone operations, no clear signal so far has been made to encourage drone operators to significantly scale up operations. In Malawi, although the Department of Civil Aviation has developed robust remote piloted aircraft regulations based on International Civil Aviation Organization guidelines, in practice, users have faced hurdles and contextual information gaps in moving toward implementation (Department of Civil Aviation et al. 2019).

The role drones will play in future transport systems will also depend on the ability of this new mode to gain the social license to grow; the key challenge will be to test the suitability of existing regulatory frameworks for mitigating possible adverse impacts while allowing drone services to develop (ITF 2018). A comprehensive review of national regulations (Jones 2017) identifies a handful of countries where outright bans or effective bans resulting from very restrictive regulations (for example, Chile, Colombia, Nigeria) are in place and no commercial flights have been approved. About 20 countries permit drone use for commercial purposes under the requirement of visual line of sight and a limit of one drone per pilot. Such regulations, in turn, have direct implications for the costs and the overall economics of drone operations. The continued accumulation of evidence on the logistics and human and societal impacts of drone operations is therefore yet more essential—the documented success of drones on a trial basis in one or two applications, as for deliveries in Rwanda, can help drive the overall acceptability of drone operations.

THE NEED TO INTEGRATE DRONES WITHIN EXISTING SYSTEMS AND SUPPLY CHAINS

The back-end work needed to ensure that drone pilot projects become long-term initiatives integrated into overall government systems should not be underestimated. In the medical use cases, it is important that drone operations connect to health sector data systems and that overall supply chains be optimized. In the mapping and risk assessment use case, drone-imagery-based maps by themselves do not constitute formal land registration; the companies or donor institutions that help collect this data must also work with government agencies on integrating the field information into official land cadasters or registration databases.

The funders of UAV pilots have an important role to play in enabling sustainable scale up. In Tanzania and Zanzibar, where drone technology has been used in a range of cases—from risk assessment to land titling, to antimalaria crop dusting, to infrastructure inspection—all of the projects have been initiated and funded by outside actors despite receiving subsequent support from local and national government entities. It is important that pilot funders be transparent about their expectations for cost reductions over time because they may be willing to pay more for a prototype today but expect much lower prices for UAV solutions at scale (USAID 2018).

THE ROLE OF SUPPORTING INFRASTRUCTURE

The economics of drone applications in the region also continue to depend on the presence of supporting infrastructure. The only fully functioning drone-based medical goods delivery system in the region, Rwanda's, relies on a well-developed cellular network (3G and 4G) given that orders are placed by text message to the distribution center (USAID 2017). For drone-based aerial image acquisition in applications such as land mapping, disaster risk assessment, and crop assessments, cloud-based image processing remains challenging because of gaps in internet connectivity. In Malawi during the dry season (mid-May to mid-November), power cuts may last six to eight hours, with a possibility of full days without electricity. The Electricity Supply Corporation of Malawi provides a blackout schedule, but it is not always adhered to (Department of Civil Aviation et al. 2019). For the country-wide manned aircraft–based mapping project undertaken by Pasco Japan and COWI Denmark in Uganda in 2015–18, production of the orthophotos required significant computing power, such that the data-processing stage of the project had to be done mainly outside the country (Noergaard, Kristensen, and Sato 2020).

REFERENCES

African Union and NEPAD (New Partnership for Africa's Development). 2018. "Drones on the Horizon: Transforming Africa's Agriculture." NEPAD, Midrand, South Africa.

CTA (Technical Centre for Agricultural and Rural Cooperation). 2019. *The Digitalisation of African Agriculture Report 2018–2019*. Wageningen, the Netherlands: CTA.

Department of Civil Aviation, MACRA, VillageReach, GIZ, and UNICEF. 2019. "Malawi Remotely Piloted Aircraft (RPA) Toolkit: A Guideline for Drone Service Providers and Implementers in the Development, Humanitarian and Research Fields." December. https://www.villagereach.org/wp-content/uploads/2019/12/Malawi-RPA-Toolkit-2019_December.pdf.

Drones for Development. 2016. "Dr.One Proof of Concept. Executive Summary." November. IDI Snowmobile, Amsterdam.

ITF (International Transport Forum). 2018. "(Un)certain Skies? Drones in the World of Tomorrow." International Transport Forum, OECD, Paris.

Jones, T. 2017. "International Commercial Drone Regulation and Drone Delivery Services." RAND Corporation, Santa Monica, CA.

Noergaard, P., O. S. Kristensen, and K. Sato. 2020. "Efficient Country-Wide Aerial Image Capture as a Foundation for Topographic Mapping, Cadastral Mapping and Land Administration Systems." Paper presented at the Annual World Bank Conference on Land and Poverty, Washington, DC, March 16–20.

UNICEF Supply Division. 2019. "Unmanned Aircraft Systems: Product Profiles and Guidance." October. UNICEF Supply Division, Copenhagen.

USAID (United States Agency for International Development). 2017. *Unmanned Aerial Vehicles Landscape Analysis: Applications in the Development Context.* Global Health Supply Chain Program-Procurement and Supply Management. Washington, DC: Chemonics International Inc. for United States Agency for International Development.

USAID (United States Agency for International Development). 2018. *UAVs in Global Health: Defining a Collective Path Forward.* Center for Accelerating Innovation and Impact (CII). Washington, DC: United States Agency for International Development.

USDOT (United States Department of Transportation. 2018. "Unmanned Aircraft System Applications in International Railroads." US Department of Transportation, Federal Railroad Administration, Washington, DC.

VillageReach. 2019. "Toolkit for Generating Evidence around the Use of Unmanned Aircraft Systems (UAS) for Medical Commodity Delivery." Version 2, November. VillageReach.